THE ARMAMENT TIDE

Rearming America

Rear Admiral Stuart Franklin Platt
Supply Corps, United States Navy, Retired

With Duffrey Sigurdson

Granville Island Publishing

Copyright © 2002 Stuart Franklin Platt

All rights reserved. No written or illustrated part of this book may be reproduced, except for brief quotations in articles or reviews, without written permission from the author or his estate.

National Library of Canada Cataloguing in Publication Data

Platt, Stuart Franklin, 1933-
The armament tide : rearming America / Stuart Franklin Platt, with Duffrey Sigurdson.

Includes index.
ISBN 1-894694-17-1

1. United States—Armed Forces—Procurement. 2. United States—Armed Forces—Appropriations and expenditures. 3. Defense contracts—Economic aspects—United States. 4. Defense industries—United States.

I. Sigurdson, Duffrey, 1957- II. Title.

UC263.P52 2002   355.6'212'0973   C2002-911215-X

Editing: Graham Hayman
Proofreading & Index: Ann McTaggart
Cover and Book Design: Rebecca Davies

First Printing November 2002
Printed in Canada

Granville Island Publishing
Suite 212 – 1656 Duranleau
Vancouver, BC, Canada V6H 3S4
Tel: (604) 688-0320 Toll Free: 1-877-688-0320
info@GranvilleIslandPublishing.com
www.GranvilleIslandPublishing.com

All photos from author's collection unless otherwise indicated

I humbly dedicate this book to the Armed Forces of the United States of America, who serve and protect this Nation on land, at sea and in the air!

Stuart F. Platt

TABLE OF CONTENTS

Foreword	VII
Chapter 1	
The Flood Tide of Rearmament	1
Chapter 2	
Big Ships, Big Costs, Big Savings	13
Chapter 3	
The General Electric F404	22
The Engine that keeps going and going and going	
Chapter 4	
Aegis – Hell hath no fury…	30
Chapter 5	
The Silent Service	40
"Underhand, underwater and damned un-English"	
Chapter 6	
Smart Weapons need Smart Buyers	53
Photos	
Chapter 7	
Honor Abides Here	64
Gettysburg and Homeland Defense	
Chapter 8	
Damn the Torpedoes	72
Congress, the Senate and Procurement	
Chapter 9	
The Crusade for Sensible Procurement	79
Today's Blunders are Tomorrow's Bills	
Chapter 10	
Charting Our Course to the Future	90

CHAPTER 11
  Greasing the Wheels on the Bandwagon of Change   102

CHAPTER 12
  War and Peace in the 21st Century   116

EPILOGUE
  Victoria Per Perseverantiam Venit   131
  Through Perseverance Comes Victory

LIST OF ACRONYMS   145

DOCUMENTS   147
  Personal Letters; Fitness Reports
  1986 Report to Congress

APPENDIX A   171
  World-wide GDP
  CVN Fact Sheet

APPENDIX B   179
  Business Principles by Chapter
  Business Principles by Topic

INDEX   200

ACKNOWLEDGEMENTS

On this page I would like to thank, pay my respects to and acknowledge the many people who have given so freely of their time and wisdom in the development of this book. Two people particularly, have been most helpful to me in this endeavor. First, Duffrey Sigurdson, who has worked with me most effectively and has a keen ability to help weave a story together. Second, William Glover, whose sharp business talent was called upon to assist in drawing up the "Business Principles by Chapter" and "Business Principles by Topic". I would also like to thank my editor Graham Hayman and my friend and publicist Barbara Sweetman.

I am also indebted to my friends, the Honorable Steven Honigman, former General Counsel of the Navy; Tom Burgess, President of Blue Stone Communications; and professor Arthur Danese, mathematician and scholar, for their insight and boundless encouragement. Among the group who helped me in the writing of this book, none was more patient or understanding than my wife, Melonee Ann Daniels.

I would also like to acknowledge the blessing I have received in my daughters, Elizabeth, Nancy, and Jennifer. Elizabeth was in Tower 1 of the World Trade Center on September 11, 2001 and we are thankful to have each other today when so many others were lost to us forever.

Stuart F. Platt

Bainbridge Island, Washington - July 22, 2002

FOREWORD
Hon. Steven S. Honigman

Our armed forces are the product of a multitude of decisions and plans. Some involve strategy, others technology or personnel. They require analysis, objectivity and vision. Underlying all of them is a common element – the inescapable economics of procurement.

Defense procurement takes place at the intersection of often incompatible principles. On the one hand, the cost of producing items such as ships, weapons systems, software or bullets will be lower if production capacity is used to the fullest, and on that basis the government can expect to spend less for the items it buys. At first glance, this may argue for awarding a complete buy to a single producer, and letting other companies' unneeded production capacity wither on the vine. But eliminating or restricting opportunities for competition ultimately drives prices upward. Even patriotic defense contractors have obligations to their shareholders and creditors to seek enhanced profits that single-source pricing can inspire.

Moreover, innovation in design, production and follow-up improvements will suffer if competitors and their technical teams do not constantly strive to push the envelope and improve the product in order to win contracts away from one another.

And in uncertain economic times, a company that the government depends upon to supply a vital part or product may fail to survive, for reasons completely unconnected to the terms of its government contract or contracts. Yet the resources the government can spend on each procurement are necessarily limited, and funds spent purely to foster competition on one defense program mean less money available to buy needed items of another kind. Where should the appropriate balance be struck?

Whether the item in question is a jet engine, a submarine, an attack bomber or an artillery piece, these conflicting imperatives are embedded in every procurement. The current defense environment – transforming the way that our armed forces will be designed to fight, anticipating and countering a host of unprecedented asymmetric threats and creating effective homeland security instantaneously – involves ever-increasing levels of complexity and risk. None of the solutions will be easy or simple, but each procurement decision will have to be right.

To this critical rearmament arena, Admiral Platt brings thoughtfulness, insight and experience. During his distinguished naval career, he perfected his skills under two formidable leaders – Admiral Hyman G. Rickover, the creator of America's nuclear navy, and Secretary John F. Lehman, the architect of President Reagan's six-hundred ship fleet. As Deputy Commander of the Naval Sea Systems Command and then as the Competition Advocate General of the Navy, Admiral Platt understood and applied the lessons of competition to strengthen the Navy's procurement of warships, aircraft and weapons systems that are at the heart of our defense capabilities today.

I served as the General Counsel of the Navy during the Clinton Administration, after Admiral Platt's tenure as the Navy's first Competition Advocate General. My colleagues and I grappled with the same fundamental issues that Admiral Platt confronted at the Pentagon and addresses in this book. Those issues are timely ones. They have enormous significance for our national defense today.

Admiral Platt's persuasive advocacy of the benefits of competition makes *The Armament Tide – Rearming America* an important voice in the national defense procurement debate. For me, two of his themes are particularly significant. First, he recognizes that fostering competition among suppliers to the government is vitally important but not enough. Equally necessary is enabling the government to compete effectively with its suppliers in negotiating defense contracts, monitoring their performance and resolving claims. To achieve that goal, Admiral Platt recommends provocative changes to procurement officers' career paths and creation of a uniformed procurement czar at the most senior level to enhance the government's firepower at the negotiating table.

Second, Admiral Platt understands the consequences of procurement failure. Among my responsibilities as the Navy's General Counsel was the defense of one of Admiral Platt's prime examples, the A-12 case. More than ten years after this attack bomber program was cancelled, it is still the subject of intense litigation which both sides can ill-afford to lose. The prize, over three and a half billion dollars plus interest, would pay for one of the aircraft carriers that Admiral Platt bought for the Navy, and the cost of the parties' legal teams would fund part of that carrier's air wing. As Admiral Platt correctly emphasizes, avoiding another A-12-like confrontation in canceling weapon systems in the future is an essential requirement for a successful rearmament program. Would the A-12 case itself have arisen if Admiral Platt's Chief of Defense Material Procurement had been on the scene to bring his judgment to bear? Would such a leader's engagement deter future A-12 cases from taking place?

In *The Armament Tide – Rearming America,* Admiral Platt raises important questions and offers thought-provoking advice. Through his central thesis that competition works to create better procurements, he makes a significant contribution to our national defense.

                              Steven Honigman, NYC, August, 2002

Appointed by President Clinton and confirmed by the Senate, the Honorable Steven S. Honigman served as the General Counsel of the Navy from 1993 to 1998. He received the Department of the Navy Distinguished Public Service Award and was recognized as a leader in acquisition reform, procurement-related litigation and the accomplishment of national security objectives in the context of environmental law compliance. He is a member of the commercial litigation and government affairs departments of Thelen Reid & Priest LLP and serves as a member of the boards of directors of a high-tech defense electronics company and the leading producer of combat rations for the United States armed forces.

CHAPTER 1

THE FLOOD TIDE OF REARMAMENT

Rearming America is a process that in many ways is similar to the action of the tide as it ebbs and flows. At the time I began to write this book, a scant six months after the September 11, 2001 attack on the World Trade Center, the rearmament tide had begun to flow strongly again and it is my hope that this book will help our nation's leaders and the American people better understand how to manage the business of rearmament during these flood tides.

Stalin's Red legacy and the resultant rise to superpower status of the Soviet bear after WWII was the all-consuming focus of our nation and our military, right from my University years through to my retirement as a Rear Admiral of the U. S. Navy in 1987 (see #1 in Documents). To counter the Russians and other global threats, the world looked to the USA. Our men and women in uniform served with pride, we armed them with the best of weapons and technology and our enemies feared our power both at sea and on the actual and potential battlefields of the world.

Sustaining that power is expensive and in many cases during my career we were paying far too much money for greatly needed systems, and we were doing so in a time of great fiscal restraint within the nation and in Congress. By 1980, it was obvious to many in the newly elected Reagan administration that something had to change within the Department of Defense

(DoD) procurement system. We were not getting enough bang for our buck. It should be acknowledged that the procurement system itself was basically sound, as were the business practices of the vast majority of defense contractors at that time. That is still true today.

However, by 1982 attitudes had started to change in the procurement world. Both the Navy Secretariat and Defense Secretary Caspar Weinberger were coming to the realization that even a 10% savings in purchasing major ships and systems would mean savings in the billions of dollars for the American taxpayer.

Historically speaking, the procurement groundwork to revitalize and prepare our military for the post-Vietnam era and on into the twenty-first century reached its zenith during the period 1983-1987. President Reagan and then Secretary of the Navy John Lehman were intent on a 600-ship Navy to counter the significant Russian threat at sea and thus on July 12, 1983 I was appointed by Lehman to work with that goal in mind. I was to serve my final years in the Navy as the first Competition Advocate General of the Navy, beginning a full year before Congress would mandate such an office for each of the branches of the Armed Forces and NASA.

The sophistication of new weapons and new systems means that major programs for new weapons, ships and aircraft take years to get from the drawing board into the hands of the military personnel who use them, and often the contracts to buy them are in the billions of dollars. The United States was about to begin a program of rearmament that would see us building some eighty new ships and a further hundred or so vessels to replace older, less capable ships already in service with the fleet. Many of the capital ships and systems were built under multi-year contracts and more often than not were sourced from a single supplier. In too many cases, competition was non-existent

THE FLOOD TIDE OF REARMAMENT

and opportunities to take advantage of the powerful force of normal market-driven efficiencies were being lost to both our suppliers and ourselves.

In fairness to the contractors, we should remember that long before America had a Navy, Britannia ruled the seas. She did so at great cost to the public purse and as historical documents show, the Admiralty was often at the mercy of its suppliers. When you are the only shop in town that can make a set of sails for a 72-gun ship, you can be pretty much assured of having the upper hand in the contracting process. It is no different today for the U.S. Navy when sourcing strategic platforms or systems where there is only one supplier.

Charging what the market will bear is one of the cornerstones of a strong capitalist system. When a monopolistic situation is allowed to develop (see chart following) this can have unfortunate consequences and strong anti-trust legislation has therefore been enacted that usually prevents any abuse of that power within the market place. Neither the Navy nor the government have any desire to inhibit the right to do business, and in fact the intent of the procurement changes that were about to occur in the U.S. Navy in the early 1980s was to encourage *other* businesses to compete in areas where we only had single source supply.

Turning for a moment from a military to a strictly business point of view, most CEOs would be alarmed by the implications of single-source supply to a company's production line. It should be no different within the military, but sadly as can be seen below, that was not the case then and is not the case today, despite our best efforts in the 1980s.

The current trend in corporate America is to merge into even larger companies, using the merger to acquire a complementary business or to swallow a rival. A look at four of the biggest defense contractors is clear illustration of that trend. All have become

3

much stronger as a result of mergers in the past decade and each of these four provides a broad array of crucial weaponry and systems.

THE MEGA CONTRACTORS ARE GETTING BIGGER

BOEING
1996   Boeing merges with Rockwell Aerospace/Defense
1997   Boeing merges with McDonnell Douglas Corp.

LOCKHEED MARTIN
1995   Lockheed merges with Martin Marietta
1996   Acquired Loral
1998   Lockheed Martin/Northrop Grumman merger blocked

GENERAL DYNAMICS
1995   Acquired Bath Iron Works
1997   Acquired Advanced Technology Systems
1999   Acquired Gulfstream Aerospace
2001   Merger with Newport News Shipbuilding blocked

NORTHROP GRUMMAN
1994   Merger of Northrop and Grumman
1996   Acquired Westinghouse Electrical Systems
1997   Acquired Logicon
2001   Acquired Litton Industries
2001   Acquired Avondale and Ingalls in Litton acquisition
2001   Acquired Newport News Shipbuilding
2002   Acquired TRW

Since the early 1990s there has been a significant consolidation within the defense industry. There are now less than ten significant defense firms, down from more than fifty competitors in the 1980s.

THE FLOOD TIDE OF REARMAMENT

In light of this changing business environment it is time to "purchase smartly and think outside the box." This is particularly true today as we see more and more mergers of defense contractors into ever-larger companies with ever-stronger influence. The blocked merger* in 1998 of Lockheed Martin and Northrop Grumman would have seen the creation of a company that would control a full 25% of the entire defense procurement budget!

Due in large part to the unique needs of the military, the Navy and other branches had often been at the mercy of sole source supply, which resulted in contract structuring that was unfavorable to the military, and in the early 1980s we were getting some bad publicity as a result. The press was having a field day, reporting on $700 wrenches and other seemingly ridiculous expenditures. Overhauling the procurement system was a job I felt I knew how to do, and with my appointment as the Competition Advocate General of the Navy, I had the support of senior members of the Administration and I looked forward to the challenge. It was, however, going to be an uphill battle and everyone involved knew it.

Having swept to power with 489 electoral votes, the Reagan administration was able to run interference in Congress and

---

*Attorney General Janet Reno stated in a Department of Justice press release dated March 23, 1998 that, "Competition is the lifeblood of a free market economy. It produces better products, innovation and lower prices for consumers. This Administration's antitrust enforcement policy has been based on a simple principle: in a fast-moving global economy, our nation cannot afford anything less than full-blooded competition.

That is why we asked a federal court in Washington, D.C. today to prevent the merger of the Lockheed Martin Corporation and the Northrop Grumman Corporation – two of the largest suppliers of defense systems to the U.S. military. This is the single largest merger ever challenged.

In our complaint we allege that the proposed merger would substantially reduce competition in many areas of vital importance to America's national defense. It would cost taxpayers more and take the competitive wind out of the sails of innovation in the production of many critical systems that protect our fighting men and women."

continued next page

THE ARMAMENT TIDE

in the Pentagon, and so the Navy re-opened for business. We had a mandate from President Reagan that would soon make Secretary of the Navy Lehman and myself two of the most unpopular men in the boardrooms of defense contractors throughout America, a sentiment that would linger for a number of years while true competition in the military / industrial relationship was instituted. Happily for us, our popularity in Congress would increase rapidly during those years due to our successes in contracting and our efforts to bring sound business practices to the contracting process.

At the time of my retirement in 1987 the first hints that the Soviet empire was collapsing upon itself were beginning to show through, but to say that our military planners had foreseen the fall of the Berlin wall would be untrue. We knew, however, that its foundations were crumbling: our intelligence sources had given us a good picture of the internal difficulties in Eastern Europe.

Senior planners in the U.S. military nevertheless expected that in any foreseeable worst case scenario, our forces would be called on to fight a large-scale war against a powerful and heavily-armed enemy. Then and now, that scale of warfare calls for a strong well-equipped Navy with both a capable submarine force and an aircraft carrier based air arm with a global reach.

The defense industry is already highly concentrated. If this merger were to go forward, America could face higher prices and lower quality in advanced tactical and strategic aircraft, airborne early warning radar systems, sonar systems, and several types of countermeasure systems that save our pilots from being shot down when they are flying in hostile skies.

This merger isn't just about dollars and cents. It's about winning wars and saving lives. That's why the investigation of this merger has been conducted jointly by both the Justice Department and the Department of Defense.

Together, we want to insure that any defense merger protects our soldiers' lives and our taxpayers' wallets. Our message is simple: If a merger does not encourage strong competition, it is not in America's interest." (Department of Justice Press Release. March 23, 1998
http://www.usdoj.gov/archive/ag/speeches/1998/0323_ag)

THE FLOOD TIDE OF REARMAMENT

Capital ships cannot be built as needed. It is an enormous undertaking to commit to the construction of these ships. The time scale is in years, not months, and the money involved is staggering. I personally negotiated the first dual contract for nuclear-powered aircraft carriers and they were in the area of +7 billion dollars for the pair (the reason they were negotiated as a pair was that we wanted to take advantage of production line economics!). If you then factor in the cost for all the aircraft, electronics and personnel needed to equip these warships, you can start to understand the complexity of purchasing for the U.S. Navy.

In order to comply with President Reagan's orders to provide for a 600-ship Navy it became imperative that we put in place a program that would deliver to the officers and sailors of the Navy and Marine Corps the ships and equipment they needed, on time and at a fair price.

The chart on the following page shows that the Navy reached a peak of 594 vessels in 1987. The 600-ship Navy, for all intents and purposes, had become a reality.

The chart following indicates the decline in surface ships that has occurred since that time. While it may be noted that our strategic roles have changed somewhat and technology has allowed us to compensate for numbers to some degree, we still have the same sized oceans to transit and the world can still be a very dangerous place.

Sadly, we have shot ourselves in the foot on occasion, as was the case with some well-intended but poorly planned procurements involving the very fleet I had helped to build.

Mark Thompson, respected Pulitzer prize winner and a dogged reporter who as far back as 1982 was haunting my offices, wrote in a *Time* magazine article: "... the Navy was helping to pay for Seawolf ($2.4 billion each): by scrapping fifteen Los Angeles-

THE ARMAMENT TIDE

U.S. NAVAL VESSELS BY TYPE 1980-1987

TYPE	9/80	9/81	9/82	9/83	9/84	9/85	9/86	9/87
Battleship	-	-	-	1	2	2	3	3
Carrier	13	12	13	13	13	13	14	14
Cruiser	26	27	27	28	29	30	32	36
Destroyer	94	91	89	71	69	69	69	69
Frigate	71	78	86	95	103	110	113	115
SSN Submarines	82	87	96	98	98	100	101	102
SSBN Submarines	40	34	33	34	35	37	39	37
Command	3	4	4	4	4	4	4	4
Mine Warfare	25	25	25	21	21	21	21	22
Patrol	3	1	4	6	6	6	6	6
Amphibious	63	61	61	59	57	58	58	59
Auxiliary	110	101	117	103	120	121	123	127
Surface Warships	191	196	202	195	203	211	217	223
Total Active	530	521	555	533	557	571	583	594

class subs this year and next, some of which spent barely half their 30-year life-spans at sea. The cost to the taxpayer: about $2.4 billion. The Navy is doing this with other vessels as well. For example, twenty-four guided-missile frigates bought during the Reagan Administration for about $7.2 billion are being retired after spending only 46% of their projected life at sea. Eight will have steamed for just 14 years of their expected 35 years of service." (Mark Thompson, "The Navy: Letting Ships Die Before They Get Old." *Time* Sept. 29, 1997)

The post of Competition Advocate General of the Navy created in 1983 was intended to oversee the implementation of a sound acquisition strategy within the Navy. Until my appointment, many defense contractors could seek the most favorable terms for themselves without threat of being displaced by a competitor. Of the $232 billion dollars approved in the 1983 defense budget, approximately $44 billion was allocated for U.S. Navy procurement alone, not including R&D and other items. Yet less than 23% of its contracts were awarded via competitive bidding. By 1986, we had gotten the figure up to more than 75% and in the initial two years of the program, we were able to show Congress a savings of +$7 billion, in other words, the equivalent cost of two aircraft carriers! Along with these savings, we also learned the benefits of using more off the shelf parts and equipment. The commercial electronics industry, for example, had come a long way in producing reliable and robust equipment and by "purchasing smartly and thinking outside the box" we were able to take advantage of these available products at great savings to the Navy.

Despite the obvious success of our efforts, our biggest battle to refine the procurement process came in the halls of Congress in 1985. Heavy opposition was still forthcoming at that time

U.S. NAVAL FORCES SEPT. 11/2001

TYPE	ACTIVE 11/2001
Battleship	-
Carrier	12
Cruiser	27
Destroyer	54
Frigate	35
SSN Submarines	54
SSBN Submarines	18
Mine Warfare	27
Patrol	13
Amphibious	39
Auxiliary	58
Surface Warships	116
Total Active	337

from within the Navy organization as well. The reason? Bureaucrats feared a loss of control of the process and a change in the status quo of procurement. Many believed that our new program applied only to the "belt buckles and beans" purchases. They rejected the notion that it would be an equally effective business practice for purchasing a major weapons system.

Having worked for six years on the H.Q. staff of Admiral Hyman G. Rickover, the 'Father of the Nuclear Navy' and the man who had designed and built the first nuclear submarine, the USS *Nautilus* (SSN 571), I already had a good understanding of the many problems we faced. For example, we had difficulties with the single supplier of the Ohio class or Trident nuclear ballistic missile submarines (SSBN). I felt this was an area of procurement and contracting where we could make our biggest impact and so I made a proposal to Secretary Lehman that the Trident SSBN would be the perfect candidate for a landmark program, to show Congress and the U.S. military that the biggest items could and should be obtained by competitive bidding.

The Electric Boat division of General Dynamics in Groton, Connecticut, was the single supplier of Trident ballistic missile submarines to the U.S. Navy and without doubt, Electric Boat were ready and able to flex their considerable muscle both in Congress and in the Pentagon. Lehman agreed with me that this should become one of the showpieces of our efforts and buoyed by his support, we set a course to institute competitive bidding for future Trident construction. Speak of rocking the boat! We were about to send a shockwave through the U.S. defense industry that only the very dull could have missed.

However, I am getting ahead of myself and although I am at risk of being accused of being a retired Senior Admiral espousing old doctrine, there is good reason to reflect now on our future procurements and the procurement process. Terrorism, other emerging threats and a much altered world stage mean our forces must now be prepared to react to a different enemy than we planned for in the early 1980s. It is still very much the case that we need our big aircraft carrier battle groups, of that we can be in no doubt. We have seen their worth on numerous occasions since the 1980s in situations where we have been able to project our power and protect our allies on a global basis, first by the presence of our fleet and then by its attendant striking power. Five acres of flight deck, anywhere in the world you need it, is a theatre Commander's dream come true.

Getting the most bang for our buck did not begin with our programs in the mid 1980s, nor did it end with my retirement in 1987. We simply took a hard look and a new approach to the problems of the day. It is essential that we continually review and renew our commitment to responsible purchasing on behalf of our forces and on behalf of the taxpayers. I would even go so far as to say that, given our internationally accepted role as a global force for peace and freedom, we also owe a responsibility to provide for and maintain this global force on behalf of three other groups.

Primary are the parents and families of our servicemen and women. Can we look them in the eye and say to them, "We have done our best to provide you with the tools you need"? Secondly, we owe it to ourselves, because rearmament is the people's business and our duty, and thirdly we have a duty of care to the innocent civilians of the world whom we stand obligated to protect. We are ready to protect them from despots and terrorists, not only with our bombs, but also from our bombs, via avoidable collateral damage. If we purchase wisely and well, and continually review this process, we will stand morally and militarily on the high ground and this is something we must never stop seeking to do.

THE ARMAMENT TIDE

The successes of our military forces in the late twentieth century and into this new century are a result of our earlier efforts to revitalize and re-focus them. Most of us who have worked within the U.S. Government and the military have now come to grips with the stark reality we will face in our war on terrorism and from other emerging global threats. As mentioned at the beginning of this chapter, this book is intended to help the public and those planning and studying current procurement methods to understand some of the past problems and solutions. Further along in the book I will spell out how we can apply this knowledge to the benefit of our country today.

CHAPTER 2

BIG SHIPS, BIG COSTS, BIG SAVINGS

I sat tonight watching a remarkable CNN feature on the nuclear-powered aircraft carrier, the USS *John C. Stennis* (CVN 74), filmed while it served on station in support of the war in Afghanistan. One segment featured an exceptional team of fighter pilots, namely Cmdr. Russ Knight and his wingman, Lt. Cmdr. Sara Joyner.

Since my retirement much has changed and much stays the same. The constants are the dedication, skill, professionalism and patriotism of our men and women in the military: our naval aviators are the world's best, bar none. The most obvious change apparent to old hands like myself has been getting used to fighter pilots with names like Sara!

Landing a plane on an aircraft carrier is akin to a controlled crash landing and one can only imagine the skill and sheer courage it takes to put that gut feeling of panic on hold and land it safely day in, day out, often in adverse weather or in the dark of night. I salute you all!

Another thing that stays the same is our need for a Navy that has sufficient ships of the right type, with the right systems, to carry out the orders of the Commander in Chief and to meet our strategic needs. These ships cannot be built on demand and it behooves the Commander in Chief to ensure that Congress has the mandate and the funding for construction of at least one large deck carrier per term. This is not a partisan issue for either political party; it is essential to the long-term security of the United States and the free world. The USS *John C. Stennis* took years of construction before it was ready to engage the enemy. To ensure we have these vessels when we need them in the years ahead requires three things: careful and accurate long term planning on the part of the Chief of Naval Operations (CNO), the will in Congress to fund them, and the strong industrial base to build them.

In the *Quadrennial Defense Review* (QDR) a dire warning is contained in the following statement: "The current force structure… was assessed across several combinations of scenarios on the basis of the new defense strategy and force sizing construct, and the capabilities of this force were judged as presenting moderate operational risk, although certain combinations of warfighting and smaller-scale contingency scenarios present high risk." (*Quadrennial Defense Review, Department of Defense*. September 30, 2001. http://www.defenselink.mil/pubs/qdr2001.pdf)

The current force structure mentioned above requires at least twelve aircraft carriers. There are currently thirteen, but three are non-nuclear and the oldest, the USS *Kittyhawk* (CV) 63, was first deployed on April 29, 1961. Only one of the three is being replaced with the soon to be commissioned USS *Ronald Reagan* (CVN 76). The yet unnamed CVN 77 will not be launched until the end of the decade. This will be the transitional ship between the old generation and the new CVNX class of aircraft carriers. I believe this schedule of construction should be accelerated to include one more CVN to be built concurrently with CVN 77. This extra ship will reduce the moderate to high operational risk discussed in the QDR.

During the flood tides of rearmament the will in Congress to fund capital ships is there, but so too is great need. In order to have the ships when we need them, we must be able to look ahead even during the ebb tides and be willing to fund construction while still respecting our limited procurement dollars. Funding for capital ships is a national debate and involves billions of dollars, which must be spread over a number of fiscal years, good and bad, to absorb the huge cost attached to such an effort.

The USS *John C. Stennis* enjoys a proud record and that record shows that she once steamed from Virginia to the Persian Gulf in a remarkable 303 hours for the 8000 nautical mile voyage. That quick response answered the call by then President George Bush Sr. for a rapid deployment of American forces to the Persian Gulf. No other military in the world has the capability to bring such a powerful force to battle in such a short period of time. This is testimony to the genius and skills of American designers, builders and workers who play a part in the creation of these mighty ships.

GENERAL CHARACTERISTICS, NIMITZ CLASS AIRCRAFT CARRIER

Builder:	Newport News Shipbuilding Co.
Power Plant:	Two nuclear reactors, four shafts
Length Overall:	1,092 feet (332.85 meters)
Flight Deck Width:	252 feet (76.8 meters)
Beam:	134 feet (40.84 meters)
Speed:	30+ knots (34.5+ miles per hour)
Aircraft:	85
Crew Total	5,680
Ship's Company:	3,200
Air Crew:	2,480
Displacement	Approx. 97,000 tons

THE ARMAMENT TIDE

AIRCRAFT CARRIERS AND HOME PORTS 2002

USS KITTY HAWK (CV 63), Yokosuka, Japan
USS CONSTELLATION (CV 64), San Diego, CA
USS ENTERPRISE (CVN 65), Norfolk, VA
USS JOHN F. KENNEDY (CV 67), Mayport FLA
USS NIMITZ (CVN 68), San Diego, CA.
USS DWIGHT D. EISENHOWER (CVN 69), Newport News, VA.
USS CARL VINSON (CVN 70), Bremerton, WA.
USS THEODORE ROOSEVELT (CVN 71), Norfolk, VA.
USS ABRAHAM LINCOLN (CVN 72), Everett, WA.
USS GEORGE WASHINGTON (CVN 73), Norfolk, VA.
USS JOHN C. STENNIS (CVN 74), San Diego, CA.
USS HARRY S. TRUMAN (CVN 75), Norfolk, VA.
USS RONALD REAGAN (CVN 76) San Diego CA. (commissioning in 2003)

Prior to the contracting and construction of the USS *John C. Stennis* I had played the leading role in the simultaneous contracting of two of our nuclear-powered aircraft carriers, the USS *Abraham Lincoln* (CVN 72) and USS *George Washington* (CVN 73). That experience gave me a broad understanding of the intricacies of contracting multi-billion dollar deals as well as the pitfalls of dealing with a politically powerful industrial base that, with the exception of component parts, was (and still is) an industry without competition.

Throughout 1982, debate had raged within the Senate and the House regarding funding for navy shipbuilding and conversion. An amendment had been proposed by Senators Levin and Hart to reduce funding for shipbuilding by $3.4 billion, striking funding for one of the two Nimitz Class carriers that were on the slate for that fiscal year. In the debate during the motion to table (97th Congress, 2nd Session, December 17, 1982, 4:59 p.m. Page S-15156 Temp Record, Vote # 440 3),

BIG SHIPS, BIG COSTS, BIG SAVINGS

it was shown by those favoring the motion that the General Accounting Office (GAO) had determined the savings made by procuring two carriers at a time was of the order of $750 million.* This made the Levin-Hart amendment, which was intended to be a cost saving measure, look rather empty.

The wording of amendment #1516 itself was so confusing as to be described in the debate as follows: "At this stage in the budgetary process, inaccuracies and contradictions such as these are a sure sign that something is awry. This Senate must not agree to such a poorly organized legislative effort." **

The motion to table was agreed 67-31 and by the end of this battle and a few other small skirmishes, we got the support of Congress and the United States Navy got the USS *Abraham Lincoln* and the USS *George Washington*.

Six months later, in June of 1983 I received a very kind and appreciative letter from the outgoing Assistant Secretary of the Navy George Sawyer (see #2 in Documents). Sawyer, along with Frank Carlucc,i was the architect of our dual purchase contracting. He was also a Yale graduate and capable submarine officer. After he left the Navy and prior to his duties as Assistant Secretary, he was an equally skilled businessperson. He brought the best of all three talents to his work as Assistant Secretary of the Navy.

The negotiation of the USS *Abraham Lincoln* and the USS *George Washington* had been a long and hard-fought process. We

---

* In contrast to the GAO estimate of $750 million in savings for the two ships, John Lehman in his autobiography published in 1988 shows an actual savings to government of over $2 billion compared to a "split buy" type of contract for the first two carriers. [pg. 250] He called it "one of the most successful and innovative ship contracts in history" [pg.171 John F. Lehman Jr, *Command of the Seas*. New York: Charles Scribner's Sons; Naval Institute Press, 1988

**Levin recovered from "such a poorly organized legislative effort" and today is the Chairman of the Armed Services Committee and still represents Michigan in the Senate. Senator Hart went on to some notoriety on the Presidential campaign trail before fading from the national political scene.

had wanted to make the terms and conditions of the contracts such that it would not be easy for the contractor to get frivolous change orders or develop other ways to gain after-the-fact profits for what should have been known in the first place. In other words, it was a + $7 billion dollar contract and we wanted it to be as comprehensive and airtight as possible.

I had been very impressed with the work of the Naval Shipyard in Bremerton, Washington, in refueling the eight reactors of the USS *Enterprise* (CVN 65), an engineering feat almost as difficult as building the ship itself. Norm Dicks, the very effective Congressman from Bremerton, was never slow to remind anyone who would listen how capable the yard in his district was. Most of the shipyard workers in Bremerton, incidentally, had signs in their front yards urging Norm's election, even in non-election years!

I liked what I saw both in the skill and dedication of the work force and in the quality of the management at that yard, so much so that I was able to convince Lehman to say on record that we felt that a competitive role in aircraft carrier construction could be found for that Naval Shipyard. This sent a loud message to Newport News Shipbuilding, the single-source supplier of aircraft carriers. We used this theme of competition for carrier construction to great effect in speeches during the early negotiations and I quickly learned that the realistic threat of competition was almost as good as the real thing. It was certainly difficult to pull off and we had a few close moments during the negotiations, but in the end it worked out our way. I was able to put the contracts in my safe and set about the task of getting them cleared for awarding by my superiors.

During these negotiations I had received strong support from Admiral Earl Fowler, Commander of Naval Sea Systems Command (NAVSEA). He was a no-nonsense executive and wanted to

BIG SHIPS, BIG COSTS, BIG SAVINGS

bring about accelerated achievement of the President's 600-ship Navy. A nuclear-powered aircraft carrier production line was essential to this objective. These ships are built in massive dry-docks, so dock availability was a key concern. Planning for two carriers at a time allowed a more streamlined and productive use of these vital facilities. Admiral Fowler had given me free reign in my role as head of contracting for shipbuilding and overhaul of Navy ships and he then gave me his full support to implement our plans. He was not only a good engineer but also a canny politician: he liked to keep me as a buffer between himself and Admiral Rickover, who controlled or had a hand in every aspect of the nuclear navy.

After briefing Admiral Fowler on the contract, I went to see Admiral Rickover. Much has been written about Admiral Rickover and much of it is true. I had worked under the man earlier in my career for over six years (see chapter 5) and I had survived and flourished in that time, so I knew him well and I was prepared in advance for the nuances of holding forth in his office. I handed him a point (briefing) paper, which he read with his usual rapid and astute grasp of weighty issues, and he immediately began asking the first of many substantive questions. He persistently questioned me on the price but I held the line that it was my belief that this was not only a fair price, but also the best we could do and that time was not really on our side.

Rickover was a man who had built the world's first nuclear submarine in five years from the ground up so he understood the statement clearly. With one of his usual abrupt dismissals he told me, "Do what you think is right, Sir, I have other matters to attend to." I was not done quite as quickly as I thought, however: even in a deal as sensitive and as large as this one, Murphy's Law had its role to play. At the last minute, I had to run back over to Rickover's offices and brief his staff in extensive detail: it seems the Admiral had lost the briefing paper I had left him. Rickover had been an Admiral longer than I had been in the Navy and his logistics staff, headed by Tim Foster, were a very capable bunch who had weathered more than a few crises, both minor and major, and they worked the issue well despite the surprises. With Rickover's second blessing it was an easy matter to get the Administration fully onside and things moved quickly from there.

I started to formalize the contract process just before Christmas and got authority from the administration to award the contract shortly before New Year's Day 1984. The news of this contract was significant: it was the biggest single contract award by the government in our history, a distinction once held by the Panama Canal. I am happy to say now and with much hindsight, that those decisions and our tough negotiating tactics have stood the test of time.

This contract award was announced coincidentally while Congress was in recess and most of Washington had returned to their home states. It was seen by some as a bold move and I spent most of the next six months defending myself and the contract, both on the Hill and at the Pentagon. In one particularly memorable session in early December of 1984, prior to the announcement, I was called before a Senate panel and thoroughly grilled by a committee chaired by Senator William V. Roth (R - Del) on cost over-runs and non-competitive contract awards. We were now about to answer to the American people for those costs.

The Secretary of Defense had sent an observer along from the Navy side who sat with other legislative affairs types, and I assumed I might be hung out to dry if I did not do well in my testimony. On the plus side, I knew the merits of the issue were

BIG SHIPS, BIG COSTS, BIG SAVINGS

with the Navy in all cases so I had little to fear provided I could present our case well. Some of these cost over-runs were a direct result of bitter lessons learned in war, in this case the Falklands conflict. We were reluctant for obvious reasons to admit publicly our concern that the Argentines, with very limited numbers of aircraft or skilled pilots, had inflicted punishing blows on the British fleet. They were using an excellent stand-off missile produced by the French, the EXOCET, and as a result, we had carefully reviewed our own fleet and added substantial armor plating in areas we felt were vulnerable to this very dangerous weapon. In the case of the carriers under construction, this added to the costs, primarily through disruption of the production schedules.

Senator Roth was cordial but tough during the session and by mid-afternoon we were through. Roth himself said at the hearing, "We've got to put the government in the position where we're not dependent on the good will of the contractor" (*Washington Post*, Dec. 6, 1984), so we were not too far off in our approach to the problem. I stopped by to see the Senator before I left the Senate Office building late that afternoon and he jokingly said to me, "You survived, Admiral."

I learned later that the DoD representative reported to his superiors that I was very polite and gave straight answers to the flood of questions. He summarized by saying that the senate staffers had lost interest when they saw that they could not create a big bump on my head and my career. It had been closer than I realized, but our Navy programs had won great respect that day in the Senate.

CHAPTER 3

THE GENERAL ELECTRIC F404
The engine that keeps on going and going

Our Navy aircraft carriers and their raison d'être would be next to useless without the superb aircraft that provide the punch and the protection that the fleet needs to carry out their missions. Our pilots will tell you (and our adversaries will reluctantly admit) that American-made fighter aircraft flown by our Naval aviators are second to none.

Fighter planes need maximum engine thrust and plenty of payload: there is no point in arriving at the party late or bringing the wrong dish. In other words, you have to get in fast and hit hard, and so for aircraft of this type, thrust to weight ratio and operational range are the baselines from which all other factors are measured. The engines are carefully matched to the expected aircraft weight, weapons load, and operational expectations. Along with the electronics package and airframe, engines are the main components of fighter aircraft. For operational reasons the Navy now uses only multi-engine fighter aircraft and although there are several good jet engines on the market, only a very few can make the grade when it comes to our front-line Naval fighter and attack aircraft.

Before a new aircraft type is far along on the drawing board, a detailed analysis of operational requirements has been done. By the actual design stage we can be sure we will have the right engine performance for the right job. The current front-line fighter in the U.S. Navy is the F/A-18 Hornet, which is being upgraded by the procurement of approximately 540 F/A-18E and F/A-18F Super Hornets, the first of which entered operational service in 2000. These remarkable aircraft are powered by GE F414 engines derived from the GE F404.

These aircraft will eventually be replaced by the F-35 JSF (or Joint Strike Fighter). Operational flying for the F-35 is still a long way off and there are many good years left for the F/A-18s and the Super Hornets as well. The lessons learned in the procurement process for powering these aircraft must be applied to the acquisition process for the F-35 in the coming years.

In 1984 General Electric had worked itself into a position where it was the sole-source supplier of engines in the F/A-18 program, then in its early days. G.E. were proving themselves a very difficult company to deal with. The Navy felt that although there had been some modest reduction in cost as production numbers had grown, it was not enough. Costs had to be reined in. Aircraft production had begun in 1979 and engine orders at the G.E. plant in Ohio for the GE F404 (as it was designated) were expected to reach a total of 3,239 units by the early 1990s. At almost $2 million per unit it is easy to see that the engine business is big business. The Navy especially is a good customer with its commitment to multi-engine aircraft and each airframe will undergo numerous engine changes in its lifetime.

Not only was G.E. the sole-source supplier of engines to the Navy but, as most people are aware, G.E. is one of the most powerful corporate entities in the world. The leverage they could bring to bear was formidable, and bring it to bear they did, both against their competitors in industry and in contracting with the military. As a result, in February of 1984, G.E. had managed to secure the contracts for 75% of the engines supplied for the USAF F-16, engines which until then had been produced by Pratt and Whitney.

Pratt and Whitney was at that time a division of United Technologies of Hartford, Connecticut, and the sole rival to G.E. in the fighter engine business. The blow had left P & W with a meager forty engines a year for the F-15 fighter. G.E. was in a very comfortable position at the top of the food chain.

This may have been a very good situation for G.E. but it was an intolerable situation for the military and particularly the Navy. It is important to note that this situation, which had hurt the Navy, had actually been considered a success by our Air Force brethren. They had benefited from their move to the G.E. engines (which had left Pratt and Whitney out in the cold) but it had an adverse effect on Navy engine procurement. This situation served to highlight the importance of open and well-used lines of communication between the Services in the area of procurement.

On the Navy side, we needed and intended to implement a program that would ensure that we had two sources for these engines, in order to reduce fleet maintenance costs, to bring about a price drop and to increase war-surge capacity. We did not expect G.E. to go along with the changes happily and we were correct in that assumption. In September of 1984, we were in a full-fledged fight with the senior management of G.E. and the long knives were starting to come out.

Before we continue our look back in time, it is interesting to note the current situation in fighter aircraft procurement. Recently Lockheed Martin won the contract to provide the nation with its next generation of fighter aircraft, the F-35 JSF. That contract was a bit of a mixed bag: good because it was competitively driven and bad because in the end it effectively removed the only other player in the industry. The loser of the JSF program was Boeing, with Lockheed Martin taking more than the lion's share of the work. It is likely that Boeing will be hard pressed to avoid being pushed out of the fighter plane business altogether.* Some observers suggest that Boeing's recent and rapid transfer of its corporate H.Q. to Chicago was a strategy to enable it to recover politically and financially from that body blow. The other side of the coin would suggest that Boeing will not give up its fighter business without a struggle and will seek to ensure support in Washington for the continuation of the Super Hornet program well into the twenty-first century.

Back in 1984, the news of our intention to seek a second source for the GE F404 broke in the influential *Defense Week* magazine on September 24, 1984. In his article, Richard Barnard told of industry observers who speculated that the leak of this news may have been "...simply to pressure General Electric". Nothing could have been further from the truth: we were dead set on seeing that engine second-sourced. Three days later, on September 27, a letter went out from NAVAIR to both G.E. and Pratt and Whitney, outlining our intentions.

James Krebs, Vice-President and General Manager of the Military and Small Commercial Engine Operations of General Electric, was on the receiving end of just such a letter from Vice Admiral J.B. Busey, Commander of NAVAIR, outlining the decision by the Secretary of the Navy to second source the F404 and requesting that Krebs attend an initial planning session on October 2.

General Electric was under a bit of a cloud at the time and one of their plants (Lynn, Massachusetts) had recently undergone an FBI investigation into kickbacks and fraud which resulted in several employees being fired and others demoted. They were also under suspension with us for their shoddy purchasing system

---

*In the past, it was Government policy to assure that the winner of large contracts like this would spin off a significant portion of the contract to the loser, to offset the huge costs associated with bidding on major systems, such as developing prototypes, R&D etc, and so our competitive industrial base has remained intact until now.

and we were then scrutinizing all subcontracts over $100,000 and were about to lower that figure to include anything over $50,000. This was not to say that the Navy intended to take advantage of the inner turmoil at G.E. but I give the details here to show the state of things at G.E. when they received the letter of September 27.

At the time I began work on the F/A-18 I had extensive experience in shipbuilding but this was to be my first foray as point man for aviation contracting on such a large scale. With the help of Admiral Busey, Ev Pyatt and the F/A-18 project team, we waded through the quagmire and I was able to gain the respect of the military aviation community as well.*

General Electric at the time had a distinct advantage over Pratt and Whitney and several other companies manufacturing jet engines, mainly due to its very good turbine blade technology. We had spent months prior to the September 27 letters in planning just how we could accomplish our goals of second-sourcing without putting undue or unfair strain on G.E. In fact, I recall a rather comical scene in my office (and into the adjacent corridor) where we had spread out the entire set of top-line drawings and technical data for the F404 and were crawling about on our hands and knees marking off in red, yellow and green which documents or data Pratt and Whitney could see. Despite the aforementioned problems with General Electric at the time, none were related to the engine itself: it was a first class performer in a field where second class means death.**

---

*For this effort and other work in aviation related procurement, I was chosen by the prestigious *Aviation Week and Space Technology* magazine for their 1985 Laurels award, to recognize our team's contributions to Aerospace. Admiral Busey later became Vice Chief of Naval Operations and later still, Under Secretary of Transportation. Sadly, he was killed in an aircraft accident and the nation was the poorer for his passing.

** See footnote following page

THE GENERAL ELECTRIC F404

In any case, one might wonder, without being privy to the details, how the Government (via the Navy) could demand from General Electric that it share its most sensitive technology with a direct competitor. To us, however, the "ownership" of the overall engine technology was never in doubt and even General Electric Vice Chairman E.E. Hood Jr acknowledged that they were "assets built over decades with Air Force, Navy, Army, NASA and G.E. money". (Hood's letter to SecNav Lehman, October 19, 1984. Author's personal papers)

It should be noted here that many defense companies at the time also enjoyed an unusual situation whereby the government was responsible for provision of tooling and the actual plant itself. In other words, the government in many cases was paying for the factory and the tooling, leaving the manufacturers free of any serious overhead or risk and with a very nice profit margin to boot.

In the case of G.E. and second-sourcing the F404, we were clear on the ownership of the technology, but we expected and got a strong negative reaction from G.E. and thus were fully prepared to fend off the salvos they fired our way throughout October of 1984. Krebs lobbied hard to kill the second-sourcing, including a visit to Under Secretary of Defence Taft* on October 17, 1984 followed by a hand-written letter to Taft underscoring the definitive nature of the second-sourcing and finishing his appeal with a call for a national debate: "We strongly urge National

---

**The original planning for the F/A-18 (including engine selection) for the Navy had been headed up by Admiral Michaelis who was a very experienced naval aviator and had once commanded the USS *Enterprise* (CVN 65), the first nuclear-powered aircraft carrier. Michaelis was the Chief of Naval Material and I served with him as his Executive Assistant. We often talked of the need for a better method of purchasing for major systems and of the need for an across-the-services purchasing system. Through him, I came to recognize that the economics of defense calls for the demand side of the equation to be equal to the supply side.

*see following page

policy debate on this issue – with G.E.'s participation." (Author's personal papers. Undated)

General Electric Vice-Chairman E.E. Hood Jr, in his letter of October 19, 1984 to Secretary Lehman warned: "…we see this as a much broader issue than the F404, one which should be discussed and debated within the Department of Defence, the Department of Justice and by the key committees in Congress". (Author's personal papers. October 19, 1984)

Amazingly, after years of the Navy trying to arrange a multi-year purchase, to which G.E. would respond with prices *higher* than those of an annual buy, Hood offered in his letter to Lehman a solution, "…I believe the real key to motivating us further lies in finding a solution to the current pattern of annual production buys". (Author's personal papers. October 19, 1984)

By early November we began to deliver technical data to Pratt and Whitney on the GE F404 and on November 7th a complete F404 was delivered to them for disassembly and reverse engineering. If we did not have the co-operation of G.E. we were fully prepared to go ahead without them! Not surprisingly, the very next day we received a letter from G.E. stating they "can find basis for agreement" (Author's personal papers) and by November 16, they had outlined to us a new, more co-operative corporate position on the second sourcing. They knew they

---

* The Taft Family are a singular lesson in American political history and have served in Government for well over a hundred years. William Howard Taft IV has spent his life in government and currently serves as Legal Advisor to the Secretary of State in the Bush administration. He is the grandson of William Howard Taft, the twenty-seventh president of the United States. Politics would appear to run deeply in the Taft family and they can trace their family tree back to Charlemagne, Alfred the Great and King Louis IV. Not many American politicians can claim *that* depth of political heritage. Robert Taft is currently Governor of Ohio and the Taft family roots there are likely the reason why W.H. was approached by Krebs and G.E. in 1984. The plant where the lion's share of work on the F404 would be done was in Ohio.

had been defeated and intended to bargain for the most favorable terms.

In the end, the second-sourcing of the G.E. F404 was considered a great success and although that particular program ended a few years later, it carried a strong message to the military-industrial complex that no cow is sacred. It still carries a strong message today and into the future about preserving and indeed expanding our military industrial base.

CHAPTER 4

AEGIS – HELL HATH NO FURY...

Since ancient times, ships have been given the persona of a woman as in "she's a fine ship" or "she was the fastest ship in the fleet." Aegis does not describe a ship but rather a complex and complete seaborne weapons system, and it takes its name from the Greek word meaning the shield or breastplate of the god Zeus. Zeus was the Father of Gods and Men including Ares, the God of War, who, when he returned to Olympus gravely wounded in battle against the Trojans, was berated by Zeus and told, "To me you are most hateful of all gods who hold Olympos." (*The Iliad* 5:889).

In more common times, Aegis has simply meant "under the protection of" and indeed that is exactly the intent of the Aegis-equipped ships of the U.S. Navy.

First conceived in the early 1960s to answer the increased threat to the fleet from supersonic aircraft and missiles, Aegis has been developed into the most formidable naval surface combatant weapons system the world has ever seen. The Aegis system, which is employed on both the Ticonderoga class Cruisers (CG) and Arleigh Burke class Destroyers (DDG), provides a daunting obstacle to any force wishing to attack our aircraft carrier battle groups and an answer to the critics of large deck carriers, critics who have claimed for decades that the big carriers are vulnerable and a ripe target for enemy forces.

AEGIS – HELL HATH NO FURY…

At the time, I was a Commander aboard the submarine tender USS *Simon Lake* (AS 33), whose deployment to the U.S. Naval Base at Rota, Spain was to support the SSBN 598/608 class of nuclear submarines, both in service with the Atlantic fleet and during their conversion to SSBN 640 class Poseidon submarines. The USS *Simon Lake* was more than a submarine tender: it was a floating shipyard, nuclear weapons depot, and supply center for the deployed submarines, and allowed them long periods on station without returning to the U.S. for normal home-port maintenance and supply.

These submarines were tasked with bottling up the enemy in the Mediterranean by closing the Straits of Gibraltar, the narrow gap between the Pillars of Hercules through which all ships must pass when exiting the Med into the Atlantic.* It was exhilarating work: Rota is a beautiful port (where incidentally, the midshipmen took great pleasure in those traditional Spanish delights of wine, women and song. Officers of course are immune to such temptations!) and moreover, we were a happy ship.

I was ordered from this duty to Pascagoula, Mississippi to oversee the contracting at the Ingalls Shipyard during construction of the new Spruance class destroyers, later slated for the Aegis program, as well as the construction of LHAs (Amphibious Assault Ships) and attack submarines. This gave me the opportunity to garner an early understanding of the Spruance class and their construction and what it would take to later turn them into a successful platform for the Aegis system. This dedicated project gave NAVSEA control over all aspects of the development and acquisition of the Aegis weapons system including computer software, hardware, combat systems, mechanical and electrical, spares, maintenance and tactical documentation.

---

*The Pillars of Hercules are represented by the Rock of Gibraltar on the Spanish shore and Mt Acha in Ceuta, on the North African shore.

It was the only way forward if the Navy was to build such an advanced weapons system in a reasonable time frame and within a reasonable budget.

What would set the Aegis-equipped ships apart from other more conventional warships of the day was its radar system. Known as the AN/SPY-1 it would be able to track and target over one hundred targets simultaneously. This system was first tested in 1973 aboard the USS *Norton Sound* (AVM 1) and the finished product was finally installed in the first proper ships of the class in 1983. These first ships were based on the designs for the Spruance class of destroyers and were given a promotion from "guided missile destroyer" to "guided missile cruiser" along the way. The first of that new designation was the USS *Ticonderoga* (CG 47), commissioned in January of 1983.

GENERAL CHARACTERISTICS, TICONDEROGA CLASS

Builders:	Ingalls Shipbuilding and Bath Iron Works
Power Plant:	4 G.E. LM 2500 gas turbine engines;
	2 Propeller shafts, 80,000 shaft horsepower
SPY-1 Radar and Combat System Integrator: Lockheed Martin.	
Length:	567 feet
Beam:	55 feet
Displacement:	9,600 tons full load
Speed:	30 plus knots
Aircraft:	Two SH-60 Sea Hawk
Cost:	approx. $1 billion each
Crew:	24 Officers, 340 Enlisted
Date Deployed:	22 January 1983 (USS Ticonderoga)

GENERAL CHARACTERISTICS, ARLEIGH BURKE CLASS

Builders: Bath Iron Works, Ingalls Shipbuilding
Power Plant: Four G.E. LM 2500-30 gas turbines;
              Two propeller shafts,
              100,000 shaft horsepower.
SPY-1 Radar and Combat System Integrator: Lockheed Martin
Length:
  Flights I and II (DDG 51-78): 505 feet (153.92 meters)
  Flight IIA (DDG 79-98): 509 feet (155.29 meters)
Beam: 59 feet (18 meters)
Displacement:
  Hulls 51 through 71: 8,315 tons
         (8,448.04 metric tons) full load
  Hulls 72 through 78: 8,400 tons
         (8,534.4 metric tons) full load
  Hulls 79 and on: 9,200 tons
         (9,347.2 metric tons) full load
Speed: in excess of 30 knots
Crew: 23 officers, 300 enlisted personnel
Deployed: July 4, 1991 (USS Arleigh Burke)

In the early days of Aegis a brilliant and aggressive Rear Admiral by name of Wayne Meyer was running a rather unobtrusive yet vital research center that was tracking all of the incoming and outgoing air traffic in the U.S. North Eastern Air corridor. Located just off the Jersey turnpike, Admiral Meyer and his team were getting lessons on how to manage a vast and heavily occupied battle space by learning to identify and track dozens of fast moving aircraft approaching a particular point from multiple directions. It was an innovative approach with very valuable lessons for Aegis.

Down in Pascagoula one of the problems we were looking

at was how to power the ship and the systems that Rear Admiral Meyer was helping to develop, particularly the AN/SPY-1 radar. Vast amounts of electrical power were required to run the equipment and it had to be supplied at constant voltage to prevent false returns from electrical interference that could mistakenly send the men to battle stations every time someone switched on the toaster oven. Being able to accurately detect a real signal from false was almost certainly going to be a matter of life or death for these ships, their every mission requiring them to be at the forefront of any engagement.

To provide this power required substantial generating capacity. The AN/SPY-1 alone drew four megawatts of power. To put that into perspective a modern radar found on a cruising yacht is usually four kilowatts (A kilowatt is 1000 watts, a megawatt is 1,000,000 watts)! On the current Aegis ships of the Arleigh Burke class this power comes from three Allison 501-K34 Gas Turbine Generator Sets (GTGS) producing 2500 kilowatts each and separated from each other by watertight bulkheads for survivability.

Another less than obvious but very critical part of the planning process for electrical and mechanical systems was heat dispersal. These powerful radars and other electronic equipment produced very large amounts of heat and getting rid of it on a ship was a huge problem both practical and tactical. The solution was complex and eventually solved to our satisfaction by converting DC or direct current power to very high AC or alternating voltage.

Having worked for Admiral Rickover previously, I was aware of his methods of dealing with ship builders and I have always suspected the Admiral of sending me to Pascagoula with the intent that I would keep an eye on his beloved attack submarines under construction there, to ensure that nothing in terms of manpower or focus was siphoned off from them to the Spruance ships!

The Spruance class destroyers were strongly supported by Admiral Zumwalt, the Chief of Naval Operations and John Warner, then Secretary of the Navy and later husband of Elizabeth Taylor. Rickover liked neither Zumwalt nor Warner and had little interest in destroyer ship progress during our phone calls, yet when I talked about the attack submarines, he wanted to know each word the shipyard President said. It seemed like a bully day for him when he heard from his other ears at the yard that I had had a shouting match with company officials. He often thought I was "too willing" to listen to their side, but I was quite effective at dealing with the Ingalls executive team on contracting matters and he was happy to have me there. I started a practice of writing him weekly detailed memos (Rickover was famous during WWII for his "Pinks," which were an endless stream of pink copied memos between him and his staff and evidence of his love for minutiae) and when I went on holiday during my time at Pascagoula his office would call to demand, "Where is Platt's memo?" It was an interesting exercise to balance my time between these very different but equally important Navy programs. Rickover gave me a glowing fitness report in May of 1974 that sums up very nicely our accomplishments while I was working at the Ingalls yard in Pascagoula. (see #3 in Documents)

It should be understood that at that time, career Navy officers would look with dread upon an assignment such as Pascagoula: they are not the glamour jobs in the Navy. Officers who followed a business career path in the Navy often found it provided little opportunity for promotion compared to those who chose a career path at sea. Beyond a small number of senior officers, there was little recognition in success, and a good chance of career disruption and perhaps destruction in failure. However, with costs for major systems entering into the tens of billions of dollars (the Aegis program cost is approximately $42.7 billion), the Navy began to see the value in rewarding and encouraging the best and the brightest to apply their talents to the business of arming the nation.

A memorandum from Admiral I.C. Kidd, Chief of Naval Material Command (NAVMAT), regarding my posting and likely written with the strong encouragement of Rickover, clearly illustrates the progressive attitude to business that was creeping in to the Navy command structure at the highest level. (see #4 in Documents)

It read in part…

"In view of the complex contracting issues and the claims situation at our major private shipyards, you and I agreed that two outstanding Supply Corps officers with proven records and extensive experience in contracting and contract administration should be assigned to take charge of the contract functions at each of our major private shipyards – Newport News, Ingalls and Electric Boat."

In another Fitness report that year by my superior, Rear Admiral E.E. Renfro III (Rickover also submitted fitness reports although I actually answered to Renfro), much of which repeated the Rickover report, Renfro stated, "The chief of Naval Material states that the acquisition process is a natural environment for the "businessman" of the Navy and deserves the finest officers the Supply Corps can offer…. This officer has all the attributes required to move to the most senior levels in the Navy. It is therefore considered that it would be in the best interests of the Navy to accelerate his promotion to the rank of Captain as early as possible." (Author's personal papers)

By January of 1975 my work at Pascagoula was coming to an end. We had done what we had set out to do and the Ingalls yard was now operating on a "leaner, meaner" basis and we felt confident that we had laid the groundwork and could move ahead with the Aegis development on the Spruance Destroyers without any delay or excessive problems in contracting with them.

I was ordered to duty on the staff of NAVMAT in mid-1975 to work under the new Chief of Naval Material, Admiral Michaelis and true to the recommendations of Admiral Renfro and Admiral Kidd, I received my promotion to Captain in October of that year. Christmas seemed to come all at once for me, because the next week I was told I had been selected as a candidate for a senior service college. The Navy was sending me back to school. Or so I thought. The career disruption mentioned in the letter by Admiral Kidd (see #3 in Documents) was about to impact my career and with hindsight, for the better!

Aegis, on the other hand, had everyone in school. Eventually the Spruance ships utilizing the Aegis technology would become the Ticonderoga class and these ships were in turn to be eventually replaced with a smaller, dedicated "Aegis" hull and "stuff that isn't even invented yet" as some wags would say when things got particularly complicated in the early planning process. The first of the Arleigh Burke ships was not launched until 1991 and through those years between 1975 with the first of the Ticonderoga class and today, the 31 ships based on the Spruance design have filled the job admirably. Notable amongst those early ships was the USS *Bunker Hill* (CG 52), which is still on duty today and was the first to receive the Aegis VLS Vertical Launching System. The VLS is the most advanced ship-borne missile launching system in the world. The VLS simultaneously supports multiple capabilities, including anti-air warfare, anti-submarine warfare, ship self-defense, strike warfare and anti-surface warfare, which these ships used to full and devastating effect during the Persian Gulf War.

By 1991 the first of the Arleigh Burke ships were coming on line and they were everything the original concept had sought to achieve. They were the first Navy surface ships to utilize shaping techniques for the hull and superstructure to reduce their detectability by enemy weapons and sensors. To ensure their survival in very hostile conditions, they were built with steel rather than aluminum in the superstructure, and over 130 tons of Kevlar plastic armor is used in critical areas for additional protection. Great attention was also paid to providing protection from chemical, biological and radiological attacks and the ships all have enhanced ability to absorb or protect against underwater shock (mines, torpedoes and underwater explosions), nuclear air blasts and the other more obvious weapons an enemy force would employ. Four G.E. LM2500 Gas Turbine engines power the ships with 100,000 SHP (shaft horsepower) available, driving the ships at over thirty-one knots. The sea-keeping ability of these ships, and thus their ability to fight in all conditions and in all oceans, was demonstrated in sea trials of the USS *Arleigh Burke* (DDG 51) when she successfully maintained thirty knots speed in thirty-five foot seas during a sixty-five knot gale.*

Notable among the many fine ships of the Arleigh Burke class are the USS *Winston Churchill* (DDG 81), named in honor of that great English warrior and the USS *Cole* (DDG 67), which came to the full attention of the world when it was struck

---

*Class designation is usually based on the first ship of the type, for example DDG 51 is the first of the class, and gave its name to all subsequent ships of the Arleigh Burke CLASS of ships. All of the Arleigh Burke class are built at one of two competing yards, Bath Iron Works in Maine and the Ingalls shipyard in Pascagoula, Mississippi. They are built in "flights" or "blocks" which allow for upgrading technology as it becomes available.

amidships by a terrorist vessel in Yemen harbor with great loss of life. The fact that terrorists could wound the world's most sophisticated warship (although tied to the dock at the time) and could then go on to wound our entire nation a short time later in an inconceivable attack, using innocent civilians on a civil jetliner as a bomb, clearly illustrates the evil abroad in the world today.

The Aegis story began when I was a young man, making my way and my name in the Navy, and the story is not finished yet. It was and is a complex, long-range plan that required strong support through many changes in administration (Nixon, Ford, Carter, Reagan, Bush, Clinton, and now George W. Bush have all been Commander in Chief during the development and deployment of the Aegis system) along with changes of scenery on the world stage. Although there have been some curves thrown at the program in Congress over the years, the vessels have been built competitively, they perform their duties to a degree that not even the original planners could have imagined and we continue to fund their development and deployment with the peace of mind that comes from knowing that this is a vessel that can proudly carry the label "Made in America."

CHAPTER 5

THE SILENT SERVICE

"Underhand, underwater and damned un-English"

At the end of the Victorian Era, America's own Navy was only a shadow of what it would become. Underhanded and damned un-English that they were, submarines were beginning to receive much attention from our Congress and the Department of the Navy of the United States. Submarines continue to play a vital role in the defense of the free world today and will continue to do so throughout the twenty-first century.

The quote at the start of this chapter is attributed to Rear Admiral Sir A.K. Wilson, the Controller of the Navy in England in 1900, a position that was akin to my assignment eighty-three years later as the Competition Advocate General of the United States Navy.*

In the hundredth year of the submarine service, I did have the honor of seeing my name engraved on the Cold War Wall of Fame at the Submarine Force Library and Museum in Groton, Connecticut. It still gives me a thrill when someone dear to me tells me they have seen my name engraved on that wall along with so many others who served with the Silent Service.

---

*Unlike our peers in the British military, American servicemen and women are not eligible for a knighthood. We did away with all that in the War of Independence! Article 1, Section 9 of the Constitution of the United States reads in part, "No title of nobility shall be granted by the United States: and no person holding any office of profit or trust under them, shall, without the consent of the Congress, accept of any present, emolument, office, or title, of any kind whatever, from any King, Prince, or foreign state".

The comments of Rear Admiral Wilson reflect the curious attitudes that can exist within a bureaucracy, as well as pointing out the danger of allowing major procurement decisions and the task of rearming a nation to be left in the hands of persons who may be ill-advised or ill-prepared.

Even as the Controller was making his thoughts known, the British navy was constructing a submarine under license from the United States. It was launched the following year as *Holland I*, now on display at the Royal Navy Submarine Museum in Gosport in the U.K. Many years later, I would be a guest of the Admiralty and give talks on submarine and other procurement issues at both the Greenwich War College and at Oxford. My hosts included the First Sea Lord, The Secretary from the Ministry of Defense, Michael Heseltine* and Peter Levene, head of procurement at the Ministry.

Prior to my arrival in London, Levene and I had had several good conversations on the situation they were facing. He told me that he was quite impressed with our work in the U.S. and hoped that we could shed light on the procurement process for his Admirals, some of whom were reluctant to change the status quo. He was most pleased when I took a stern and convincing approach to the procurement situation in the U.K. and the next day I dined with him and a number of the Lords Justice at Old Bailey. Talk about a stern bunch.

The United States had been involved in submarine design

---

* The day I was to meet with him in his offices, Heseltine was front and center in the morning papers. When I mentioned that I had seen the papers he in turn was courteous enough to point out to me that my visit to the U.K. had made it into the center section of the Daily Telegraph! Heseltine had some very unusual nicknames, he was variously known as "Tarzan," "Hezza" and "The Blonde Bombshell" an apparent reference to his good looks. I am uncertain as to the Tarzan or Hezza references, and no judge of men's good looks, but they are no doubt an interesting story. He was and is a colorful man to suit his colorful nicknames. He is also referred to as the "Political Assassin of Margaret Thatcher."

THE ARMAMENT TIDE

as early as 1861 with the building of the *Alligator* which entered service on June 13, 1862. It was not until the late 1880s that things began to heat up and John Holland won his first contract from the Navy. Due to the divided opinions of the time the contract was let, cancelled, let again the next year and cancelled yet again. Holland became a draftsman to make ends meet. Finally in 1895 he built his first prototype under contract, christened *Plunger*, which would seem to be a rather unfortunate name for a submarine. Despite his objections the Navy demanded that Holland install a steam engine and true to Holland's warnings, it was totally unsuited to a submarine. Engine room temperature was over 130 degrees Fahrenheit before full power was even reached. Nonetheless, Congress authorized three of these human boilers. Although a total disaster in many respects, they were invaluable as test beds for submarines to come (and for an industry, which had to learn new methods and skills). History seems to show it was a good job of contracting too: the first boat cost the U.S. Government $200,000 and the subsequent three were brought in at $175,000 apiece.

Holland had some unusual friends and backers in his career, including the Fenians (a precursor to the Irish Republican Army or IRA), who commissioned his first studies in the 1870s and then in 1881 commissioned the *Fenian Ram* which could operate at depths of sixty feet for up to an hour. It would seem that the Fenians and Holland had a falling out: the Fenians stole the ship from under Holland's nose and hid it in a shed in New Haven, Connecticut for thirty-five years. This vessel is now on display in his adopted hometown of Patterson, New Jersey.

Almost simultaneous to the development of *Plunger*, Holland was going ahead with a better design and it was at this busy period, while developing the *Holland IV*, that Holland, frustrated and broke, joined forces with a wealthy industrialist to form the Electric Boat Company in 1898. He then went on to win the first contract for a practical and effective submarine the USS *Holland* (SS1), from the U.S. Government in 1900 for the princely sum of $150,000. He won the contract in competition with his colleague Simon Lake, after whom was named the USS *Simon Lake* and on which ship I proudly served (see Chapter 4).

The USS *Holland* was driven by a gasoline engine and, it being a poor second to diesel fuel in terms of safety, Electric Boat Company was given the first contracts to produce a diesel powered submarine in 1909 and four of these were built at the Union Iron Works in San Francisco. It marked the beginnings of a long and tumultuous relationship between the U.S. Navy and the Electric Boat Company, certainly one of the great love-hate relationships in the history of the military and its contractors. Holland had left Electric Boat in 1904 and again set up in business on his own only to find that Electric Boat owned all his patents! Holland faded from the scene and E.B. was on its way.

One cannot underestimate the value of patent rights and this is a lesson still being painfully learned by the military today. Software ownership is of particular concern and contractors today can use ownership of software and data to maintain a position of sole-source supplier. They can do so because at times our Government hopes to save money by not buying in to the software or data rights in the first place. At times they are also unwilling to go to the effort required in the initial contracting process, to define who owns what in the development of systems. Not paying for or not making the effort to acquire the software or data rights can also be a pork barrel maneuver to keep favored contractors in a position of strength. We must be extremely vigilant and make it the exception rather than the rule that the contractor owns the data or software. This may cause a great many crocodile tears in Washington and corporate America but the contractors know that we are, from a business point of view, absolutely right.

Controlling software and data rights is essential, particularly so if we are forced to ramp up production while on a war footing. We will not have the time for lengthy negotiation with contractors seeking to defend "proprietary knowledge."

One of the vessels from the early buy of the Plunger series was USS A-1 (SS-2). It had been laid in May of 1901 as the USS *Plunger*. It was on this vessel in 1905 that Teddy Roosevelt became the first U.S. President to voyage underwater in a submarine. Roosevelt had been an early supporter of submarines and was so taken with the voyage and the skill and daring of the crew, that he implemented a special pay increase for the officers and men of the Navy while serving on submarines. That pay supplement is still in effect today.

Simon Lake in particular remains a favorite historical character of mine and I often reminisce while listening to Offenbach and his magnificent piece *Orpheus in the Underworld*. Orpheus was one of the Argonauts sailing with Jason in search of the Golden Fleece and he saved their ship, the *Argo*, from the Sirens with his music. Perhaps Lake shared my fondness for that music, as he named his early boats *Argonaut Jr.* and *Argonaut I*. Jules Verne, whose remarkable foresight in his classic work *2000 Leagues Under the Sea* was about to become reality, sent a telegram to Lake congratulating him. "The conspicuous success of submarine navigation in the United States will push on underwater navigation all over the world". (*Simon Lake Project* http://www.simon-lake.com/jvcable.htm)

One last historic footnote is the rather unusual tale of the submarine H-3 which ran aground December 16, 1916, in Humboldt Bay, California. The Navy tried for days to dislodge the trapped submarine using a Navy cruiser, USS *Milwaukee* (C-21), which failed despite its size and power and was itself eventually trapped in the shallow water. Eventually a group of loggers was brought in to make an attempt to free the submarine. They saw the unusual sight of a submarine aground as nothing much more than a rather large redwood on the beach and, treating the poor H-3 as such, promptly hauled it off the shore and into deep water. Legend has it that they were a long time waiting to be paid for their heroism and effort. The Naval Shipyard at Bremerton refused to pay them as the loggers did not have a written contract! The USS *Milwaukee* was not so lucky; she remained aground and eventually broke up in a storm. The hulk of the ship lies just off Samoa Beach to this day (*Dictionary of American Naval Fighting Ships*, Government Printing Office. ASIN: 9998843790; 1987).

Many years later the Bremerton Yard would launch the USS *Simon Lake* and a decade after her launch, I was aboard as we sailed from Charleston, South Carolina, to the U.S. Naval Base in Rota, Spain to relieve the submarine tender USS *Holland* (AS 32). It must have been a funny sight for the citizens of Rota to see us come in to port, the fantail (aft deck) of the USS *Simon Lake* having been used for storing the personal vehicles of the officers and crew. The ship bore a striking resemblance to a floating used-car dealership. Young sailors driving giant American-made automobiles on the narrow streets in Spain was also a remarkable sight and a story for another day.

My own acquaintance with the underwater world began in the early '60s when I spent two years on the Headquarters Staff of Admiral Rickover in Washington, D.C. While there, I learned about submarines and the purchasing and contracting of the myriad of components required for these ships, which were and still are the most technologically advanced machines on the earth or under the sea.

THE ARMAMENT TIDE

I went from there to Pittsburgh and the Bettis labs where I served as the Officer in Charge of the Reactor Engineering School, the elite training center where Rickover schooled his engineers. In that role I was intimately involved with the programs on the S5W reactor that powered the Sturgeon class of Submarines, as well as other research work on larger, more powerful reactors for aircraft carriers. At that time Rickover was already locked in battle with the Navy to carry on with a new generation of reactor called the S6G.* This reactor would rely on natural circulation for its coolant flow and thus allowed the engineers in their design work to reduce the need for many of the pumps and ancillaries that produce that greatest of enemies of a submarine, noise.

I also monitored the administrative activities of the AEC Bettis Atomic Power Laboratory, and as a result, I again caught the eye of Rickover and was sent off to Washington to head up his budget office at the AEC (Atomic Energy Commission), where I learned the insider's view of the intricacies of Nuclear Reactor procurement as it applies to submarines. I had by this point been from one end of the University of Rickover to the other and had finished my education there as I had started it, on the job. They were intense years, clouded by the race with the Soviets, who we knew from intelligence sources had submarines that were at least as fast as our own. We were able to bring the considerable resources of many capable contractors to bear when it came to researching and developing the reactors needed to meet the new threats from the Soviets and the ever-expanding role of Soviet intelligence services.

Seeing the potential for nuclear power in military and civil-

---

*Reactors were named by a sequence to identify them: in S5W, which was one of the programs I worked with, S refers to submarine, 5 refers to the model number and W signifies Westinghouse. D2G would tell you that the reactor was built for destroyers such as the USS *Bainbridge* (CGN 25), that it was the second of that model and built by G.E."

ian applications, both Westinghouse and General Electric fought furiously for contracts to develop not just reactors but component parts as well and this went a very long way to bringing down the costs of this new technology. We used the tactic (and still do to a degree) of pitting contractors against each other. In effect, we play the role of hall monitor to enforce discipline in the contracting and competition process and are thus able to keep the competitors at each other's throats more often than not. It is still an effective tactic for completing a large weapons system today and we must always strive to keep our defense industry competitive and keen.

To fall a generation behind in research and development because Congress or the military delays a program due to the deferral of costs is an unacceptable and fatal policy. Leaving a system in limbo while awaiting more favorable economic conditions or stretching out programs to pass over costs to another fiscal year and then keeping our best and brightest operating at half speed is not the way to build efficiently. We *have* to continue to find ways to make our major procurements affordable and palatable to the Government. Submarine construction and procurement is a good example of an extremely costly and technologically difficult system and crucial to our defense. Can we really afford to wait? It is folly to delay or defer on these decisions.

From my vantage point, having spent a good portion of my career directly involved with submarines and their deployment and procurement, I can say that Admiral Skip Bowman has done a splendid job of directing the Naval Reactor branch in the transition to the twenty-first century. Admiral Bowman has led the charge to the future for our submarine force, which has in the past been a victim of poor Congressional decision making and poor procurement contracting. He has shown

THE ARMAMENT TIDE

leadership and determination in getting the ship back on course. Bowman was also a graduate of the "University of Rickover" and his resumé speaks volumes about the capability of our officers and crew in the Silent Service, as well as the success of the system which trained him.

Had Rickover been less forceful in those early days than he was, it could have been that we would have lost irrecoverable time to the Soviet research effort and it is not a stretch to say that the world might well have been a different place today as a result.

For those in the know, who might be curious as to how I remember my years with Rickover, I can only say that from my perspective, the time spent under him was a great challenge and they were very productive years. I left Rickover's staff with his respect and with his kind words to the Chief of my Corps. (see #5 in Documents)

One of my clearest and perhaps saddest memories of the Admiral was at the launch of the Los Angeles class attack submarine the USS *Hyman G. Rickover* (SSN 709) in August of 1983. I flew up to Groton for the ceremony with a goodly number of Navy brass and others, but most conspicuous by their absence were the Secretary of Defense Caspar W. Weinberger and the Secretary of the Navy John F. Lehman Jr. It was a petty slight by men who should have risen above it, although their presence would likely have created an even more surreal atmosphere as both men had been instrumental in the ousting of Rickover from the Navy.* Admiral James Watkins, Chief of Naval Operations, and Admiral Kinnaird McKee represented the Navy at the commissioning ceremony. Rickover, with his lovely second wife Eleanore at his side to christen the ship,

---

*Admiral Rickover's tirade in the Oval Office when told to retire is the stuff of legend and is recounted in Lehman's book mentioned in Chapter 2.

was shunned on the platform and at the reception later, by some who were there purportedly to honor him. Lip service was paid to him by executives from both General Dynamics and their Electric Boat Division, but one had the sense that the toasts flowed from a poisoned chalice. Rickover was up to the challenge however: during his remarks he reminded the assembled "great and grateful" of the very good relationship that he and the Navy had enjoyed with Electric Boat until it was taken over by General Dynamics. He also made mention of an unnamed Electric Boat employee. Everyone knew this was a reference to Takis Veliotis, an Electric Boat manager who was hiding out in Greece in the face of an indictment on kickbacks involving General Dynamics and Electric Boat.

Following the commissioning, the reception was held at Branford House (a beautiful manor house in the English style, on the campus of the University of Connecticut). There my wife and I had the pleasure of sitting with members of Eleanore's family. During the luncheon, I made it a point to convey to them my own respect for Rickover and my admiration for what he had accomplished. Eleanore Rickover was a very gracious sponsor and in her remarks at the reception, she paid homage to the usual cast of characters and then added a twist of her own. She went on to pay tribute to "the Navy Wife" and specifically made mention of "the first Mrs. Rickover" in that salute.*

To wind up this chapter on submarines, I must admit I had a special purpose in how I have tacked through the subject. The Naval Nuclear Propulsion Program (Naval Reactors to the insiders) is a national success of global magnitude, much of this due to Rickover and some of it despite him.

I wanted to get us through a broad history so we can reflect on what was right. The program stuck to basics. We avoided

---

*Ruth Masters Rickover, 1903-1972

THE ARMAMENT TIDE

frills. Only the very best of civilian engineers and Naval officers were wanted. Long-term assignments and commitment were taken for granted. We worked to get things right and to get it simple: hardware simple enough that sailors with little more than a high school diploma and some excellent training at Navy schools could carry out their duties even during the most intense moments of battle.

Those of us who were privileged to work in this growth industry cultivated within the Navy knew our work was in the national spotlight. Money, like water, flows where it is most easily able to go. The Naval Nuclear Propulsion Program (NR) helped to make that money easier to acquire by our willingness to face Congress, the public and our contemporaries on a no-nonsense basis. We worked to make deadlines and beat budgets. We took criticism from the boss and our friends and dealt it out when needed in our own realms of responsibility. Like trained professionals everywhere, we accepted and evaluated advice, no matter in what form, style or spirit it was given.

Our attitude then was that we would always seek to find the correct technical solution, not the easiest or the cheapest; to get our contractors to make a commitment and then hold them to it; and lastly to earn the respect needed for our program by conducting ourselves accordingly. This attitude would still seem to prevail in the Naval Nuclear Propulsion Program and within the submarine force leadership today. As concerned Americans, it is our responsibility to keep that good work moving ahead. This is an area of defense where we enjoy a very substantial technological advantage over our potential enemies in the world today and it is imperative that we keep that edge. In the following charts, you can judge for yourself how far we have come. They are the impressive results of a hundred years of good headway.

GENERAL CHARACTERISTICS: HOLLAND CLASS

Builders:	Holland Torpedo Boat Company
Power Plant:	45 H.P. "Otto" Gasoline Engine
Length:	53' 10"
Beam:	10' 3"
Displacement:	74 tons
Speed:	6-8 knots
Crew:	6
Armament:	One tube, 3 torpedoes, bow gun.
Date Deployed:	12 October 1900
Ships:	7 "A" type

GENERAL CHARACTERISTICS, OHIO CLASS

Builders:	General Dynamics Electric Boat Division.
Power Plant:	One nuclear reactor, one shaft
Length:	560 feet (170.69 meters)
Beam:	42 feet (12.8 meters)
Displacement:	16,764 tons surfaced; 18,750 tons submerged
Speed:	20+ knots (23+ miles per hour, 36.8+ k.p.h.)
Crew:	15 Officers, 140 Enlisted
Armament:	24 tubes for Trident I and II, MK-48 torpedoes, four torpedo tubes.
Date Deployed:	November 11, 1981 (USS Ohio)
Ships:	18 in service

It should be noted that the Ohio class will eventually become overly expensive to maintain and there are no apparent plans to replace our ballistic missile submarines. With the demise of the Soviet Union and as a result of treaties, the U.S. has significantly reduced the number of strategic warheads we have. Of the remaining, a substantial number are still carried aboard the eighteen Ohio class submarines. Their principal role is strategic deterrence and nowadays, although they do not have specific targeting data programmed into the missiles, that data is immediately available should the unthinkable occur. Targeting the missiles, while it certainly is rocket science, isn't any more difficult or time-consuming than rebooting your computer.

A new series of submarines is now being built to replace the Los Angeles class attack submarines. Called the Virginia class, it remains to be seen if Congress and the Navy will continue to develop these much-needed but very costly vessels. The Virginia class were not even the first choice for this role: in the final years of the twentieth century, funding was cut for the Seawolf Class of attack submarines after just three ships of the projected fleet of twenty-nine had been built (see Chapter 1).

The scrapping of the Seawolf program can only be seen as a disaster of poor planning and politically based decision-making, which echoes the oft-cancelled submarines of John Holland. The demise of the Seawolf program has had a significant impact on the new Virginia class, which will take over the role of the Los Angeles class submarines. In an unusual move, Electric Boat and Newport News are "sharing" the construction of these vessels, in part due to the low production rates.

Regardless of the reasons, this method of sharing construction between "competitors" will result in greater cost per ship than need be the case. True competition is not only about price, it is also about innovation and development of technology. Sharing the contracts for the construction of the Virginia class submarines will serve to stifle the innovative thinking that comes with true competition.

CHAPTER 6

SMART WEAPONS NEED SMART BUYERS

On many occasions in my career (sometimes daily), I had opportunity to meet and work closely with the Assistant Secretary of the Navy. I worked most closely under Ev Pyatt and George Sawyer. Just outside the offices of the Assistant Secretary is a corridor with a long row of portraits and photographs of predecessor Assistant Secretaries. As portraits, nothing seems remarkable at first glance and yet they represent almost two centuries of Navy history. Some had beards, some had moustaches, some were old and some were young. Two of them, Theodore and Franklin Roosevelt, would go on to become Presidents of the U.S.A.

There is a similar gallery of portraits in the Pentagon, of all the past Secretaries of the Navy. As portraits, they are no more and no less remarkable than those of the aforementioned Assistant Secretaries. Unlike the Assistants, however, no Secretary of the Navy has ever gone on to become President and one in particular stands out for his appallingly bad luck. Secretary of the Navy Thomas Gilmer was an unwitting victim of one of the Navy's first forays into "smart weapons". On February 28, 1844, Gilmer and other dignitaries were aboard the USS *Princeton*. Built at the Philadelphia Navy Yard the year before, she was fitted with two newly designed heavy guns. One of these guns exploded during a demonstration, killing Secretary of the Navy Gilmer and several others.

As the unfortunate SecNav found out, ordnance is a

dangerous business, especially in the development stage. A "let's see if this thing works" approach takes on a completely new meaning when you are talking missiles, bombs or big guns.

The introduction of heavier guns with longer ranges, and later of "rifling" to the barrels in the late 1850s, would greatly improve our arsenal and thus our ability to impose a blockade on the Confederate Forces during the Civil War. The "smart" technology of rifling the barrel allowed the round to spin in flight and as a result it gained accuracy and range. Picture a beautifully thrown touchdown pass, then think of that same play with a wobbly pass that does not have a nice spiral to it. That is the difference between a round fired from a rifled barrel and one fired from a smooth-bore gun.

I have often wondered if it is a source of professional frustration for weapons designers that the delivery system gets more attention than the weapons they spend their careers developing. The aircraft or the ship that fires the weapon gets the press most days. It is only when something goes horribly wrong, such as the so-called "friendly fire" incidents (many of them involving "smart" weapons) or even the cannon explosion which killed Secretary Gilmer, that they come under intense public and internal scrutiny. I would hazard a guess that many Americans could tell you what *Enola Gay* was, but not near as many could tell you what *Little Boy* was.*

Military air shows and static displays are another good example. They attract huge crowds and I suspect that most of the crowd would not be there if you took away the planes and only left the missiles. Yet every ship and every aviator in the military, every sailor, every soldier and every marine, would be restricted to the role of observation post or scout without a weapon to shoot at the enemy.

---

*Enola Gay was the aircraft that dropped *Little Boy*, the first atomic bomb, on Hiroshima, Japan, in WWII.

SMART WEAPONS NEED SMART BUYERS

So far, most of the systems we have looked at have been in the billions of dollars per contract. The quantities purchased can be as low as one every three years or more and they have a life span of twenty or thirty years. It is easy to grasp that even a one percent savings on a multi-billion dollar contract is a savings worth pursuing. "Smart weapons" are another matter: they are inexpensive when compared to the plane or ship that carries them ($18,000 for a simple laser guided bomb, about $380,000 for a simple air-to-air missile). At first glance one might wonder just how much can you save on an $18,000 bomb or even a $380,000 missile? Quite a bit as it turns out, when you buy hundreds or thousands at a time, and on some days shoot them off just as rapidly.

My own experience with serious weaponry came in the early part of my career as it does for most young ensigns, who amongst their many duties will also be called to battle stations on a ship at sea. My introduction to big guns began on USS *Waldron* (DD 699).* During battle stations, my job was to help man the forward twin 5" guns. As a Division Officer, I rated a bunk in a two-man stateroom and my room mate was the ship's chaplain, an officer by name of John O'Conner, a scholarly Irish-American who went on to become the Archbishop of New York. We pounded away on the 5" guns regularly as part of the ship's routine training and I remember most vividly the complaints

---

*The 5" guns on the USS *Waldron* have been fired in anger and with great effect. She was launched and went to war during the closing months of WWII and saw heavy action then. One memorable fight took place at what was then called Cam Ranh Bay in Indochina where Task Force 38 sank forty-four ships. In a later action, at very close quarters and unable to bring the guns to bear, the USS Waldron rammed a Japanese picket ship, cutting it nearly in half and likely saving USS Dortch (DD 670) in the process. Many years would pass, and many men would serve on her, including myself, before the USS *Waldron* would be back in Vietnam. Having gone through numerous improvements and upgrades over her twenty-nine year career, she was eventually sold to the Columbian Navy in 1973. She had earned four Battle Stars from WWII and one from Vietnam.

from O'Conner that every time we fired the guns, the high pitched sound of the firing would break the bulbs of the lights in our quarters, making it difficult for him to read. For some reason only the shockwave from the forward pair of guns seemed to blow the bulbs. Our quarters were on the deck directly below and slightly aft of them and perhaps the sympathetic vibration of bulkheads and overheads whenever that gun fired was the cause. I solved the problem when we arrived in Europe by purchasing some better-quality bulbs made of thicker glass and installing them in our quarters!

One of the biggest lessons I learned from serving on the USS *Waldron* is that it is better to give than to receive, and this was something I learned from being a gunner and from the good efforts of Cardinal O'Conner! He was a pious man and a fine shipmate. I was saddened to read of his passing in May of 2000.

On one memorable live-firing exercise while serving on the USS *Waldron*, we were in company with the battleship USS *Iowa* (BB 61), and it was a remarkable sight when she fired off her big 16" guns (see photos). Unlike the high-pitched crack of the 5" gun, the 16" guns on the USS *Iowa* produced a very low but loud rumble. The projectile moved so slowly that you could actually see it, a Volkswagen-sized shell that could travel twenty-five miles and destroy an area the equivalent of three football fields. For its time and purpose it was an excellent stand-off weapon which allowed the ship to engage the enemy while remaining beyond the reach of most of that enemy's seaborne weapons. Decisive firepower indeed!

I continued to have a working relationship with the USS *Iowa* many years later when it was re-commissioned in the build-up to the 600-ship Navy (see Chapter 1) and I am still pleased to be involved with that great battleship as I write. I now serve as the Chairman and CEO of the Historic Ships Foundation in

SMART WEAPONS NEED SMART BUYERS

San Francisco, whose mission is to support the preservation and public display of this piece of Naval history. I encourage everyone to visit the ship when in the Bay area.

True "smart weapons," which are now familiar to most people via television, are the pointy end of the spear these days. They allow our Navy ships and aviators as well as the Air Force to support our ground troops very closely and to target specific bunkers and vehicles with great accuracy. Yet few of the public are familiar with just how it is that they become smart in the first place.

The answer is simple and complex at the same time. We have spent billions developing smart weapons because we have one simple need: to be able to deliver devastating firepower from a stand-off distance in order to reduce the risk to our aviators, marines and sailors who go into harm's way. Complex because in order to get the missile to do that job it has to have the equivalent of an entire aircraft radar and guidance system mounted in its nose. Complex because a stand-off weapon needs fuel to get it to the target, and thus a motor and fuel storage. It also needs something to make a decent bang, and lastly a computer and control system to co-ordinate it all. It's quite a lot to fit into a tube measuring seven inches in diameter and twelve feet long, such as the AAMRAM AIM-120 medium-range, air-to-air missile.

Just to clear the air (no pun intended) I include a simple summary of the types of missiles found in the arsenal of the U.S. military. This chart is by no means inclusive and does not include man-portable as well as many other missile systems in use today by our military.

THE ARMAMENT TIDE

Type of Missile	Primary Mission	Examples
Air-to-Air	*To attack enemy aircraft*	AIM-9
		AIM-120
		AIM-54
Air-to-Ground	*To attack enemy ground positions*	AGM-84
		AGM-88
		AGM-119B
		AGM-158
		Tomahawk
Surface-to-Air	*To attack enemy aircraft/missiles*	SM-1
		SM-2
		Patriot
		Hawk
Surface-to-Surface & Sub-Surface-to-Surface *To attack enemy land/water based assets*		Tomahawk
		Mk48 Torpedo
		Mk50 Torpedo
		Mk54 Torpedo

Surface-to-air missiles (SAMs) were progressing rapidly during my time on active combat duty. However, they tended to be pretty big weapons, fired from a huge fixed launcher and controlled by ground or ship-based radar and they could often be fooled with electronic jamming and/or chaff, which is dispersed by, and imitates the heat signature of, the aircraft under threat. As well, their range was limited to the immediate vicinity of the launcher albeit up to a pretty fair altitude. Navy pilots in Vietnam described the SAMs as "like a flying telephone pole, coming up at you." They were a very real threat to our pilots and yet they were nothing like the danger presented by the cheap, shoulder-held surface-to-air missiles that are a constant threat on the battlefield today.

Air-to-air combat was relatively rare in Vietnam and most of

SMART WEAPONS NEED SMART BUYERS

our downed pilots were lost to the SAMs and other ground fire. I stress the "relative" rarity of air-to-air combat, and this was probably a good thing considering air-to-air missiles at that time had an effective range of less than two miles on a good day, but otherwise…. Otherwise meant you then came up against the enemy in his own backyard. The Aim 7E-2 dog-fighting missile was developed as a response to the poor performance of earlier versions that had proved ineffective against the highly maneuverable MIG 21's flown by the North Vietnamese. Congressional representative "Duke" Cunningham is one of two Navy aces from Vietnam (five or more kills) and he was the first of the five American aces of that war.

Military planners realized that the unpleasant prospect of close-in air-to-air combat could be avoided altogether if an air-to-air missile were to be developed that could be fired accurately from stand-off range (outside the range of the enemy's defensive and offensive assets). The development of smarter, more lethal missiles would go a long way towards increasing the survivability of our precious pilots and their aircraft from enemy fighters. Air-to-ground missiles were also required that would give our pilots the opportunity to destroy enemy ground targets with an accurate weapon, fired from great range. Our pilots could then turn for home or engage other targets without coming into range of enemy weapons, and well before a superior opposing force could be mustered to engage them in aerial combat.

Missile technology made a quantum leap in the years between Vietnam and the end of the twentieth century. We now have a large arsenal of every type of missile, including sub-surface, sea, land and air-launched cruise missiles right on down to man-portable, shoulder-fired weapons that can bring down enemy aircraft and attack helicopters.

Close-range weapons, like the Phalanx Close-In Weapons System used aboard Navy ships, can fire 4,500 of their 20mm rounds per minute. A new Enhanced Lethality cartridge with a heavier penetrator was introduced for this weapon in the Block 1b purchase and first installed aboard the USS *Underwood* (FFG 36) in 1999. It should be noted though that few weapons can realistically fire off as many rounds as the manufacturer will claim because the barrel will simply melt.*

As mentioned earlier, small-ticket items used in great quantity should provide opportunities for substantial savings. Purchasing bullets for the Phalanx should make anybody eligible for a bulk buyer discount! This system is based on the M61-A1 Gatling Gun, which is also found in the inventory of the other branches of the service, and cross-pollination of contracts between the services for ammunition is an effective way of cutting the per-unit cost.

Although it did not occupy the lion's share of my time while I was Competition Advocate General of the Navy, I did have occasion to use the power of the marketplace to seek out the best deals for small-ticket items such as ammunition and missiles. Like many of our purchases at that time, we were faced with a situation whereby much needed munitions could be sourced from only one manufacturer, leaving us once again at the mercy of a sole-source supply line.

---

*Modern gun barrels are usually plated with hard chrome and this helps preserve the barrel during heavy fire. Unfortunately chrome imposes some performance limitations which lead to the degradation of the barrel and hence the need for frequent (expensive) replacement. These types of barrels and coatings will not be capable of firing the new projectiles being developed because they use propellants with a much higher energy output than current rounds. The Office of Naval Research (ONR) is investigating other coatings to produce the next generation of gun barrels. This technology incidentally is able to cross over to the commercial coating market, a crossover that is imperative to ensure the long-term health of small defense contractors.

SMART WEAPONS NEED SMART BUYERS

For that reason and others, I traveled to the Naval Weapons Center at China Lake, California, in 1984 to speak with the staff there about opportunities for ensuring second-sourcing of specific weapons. The China Lake folks had done a superb job of producing documentation packages on both the Sidewinder and the Sparrow missiles and we intended to have them produce the "secker" components of the HARM AGM-88, a high-speed anti-radiation missile. Although there was much protest both in the Navy and in Congress against a manufacturing role for China Lake (particularly from representatives seeking to protect manufacturers in their home states), we wanted to move forward quickly with procurement for not only the HARM missile but also for improvements to the MK 48 and MK 50 torpedoes as well.

We were also seeking, at that time, to reduce the "fly away" cost of two other major missile programs: the Harpoon AGM-84 anti-ship missile produced by McDonnell Douglas (now Boeing) and the Phoenix AIM-54C air-to-air missile produced by Hughes Aircraft Co. (now Raytheon). "Fly away" costs refer to the cost of the missile at the point where it is ready to be used in anger.

Our goal was to bring down the costs of both missiles. The Harpoon was costing us approximately $933,000 per unit and the Phoenix about $950,000 per unit. Our goal was a flyaway cost of $765,000 for the Harpoon and $450,000 for the Phoenix. Both programs were a success and today, with technological improvement over those early models, the Harpoon costs our government $474,609 and the Phoenix costs $477,131 per unit. (Federation of American Scientists, http://www.fas.org/man/dod-101/sys/smart.htm)

Hughes Aircraft Company and Raytheon initially produced the Phoenix missile. Hughes Aircraft Company was purchased by General Motors in 1985, bought out General Dynamics missile division in 1992 and finally merged with Raytheon in a $21 billion deal in 1997. It remains to be seen if such savings will be possible in the future with the ever-shrinking manufacturing base in the aerospace and defense industries.

It is worth noting that in our desire to have the best weapons at the best possible price we have done something a bit unusual for the U.S. in weapons procurement. We shop around, not just in our own neighborhood but also overseas. Both the Penguin AGM-119B, a Norwegian-designed, helicopter-launched, anti-ship missile, and the Israeli-designed Have Nap AGM-142 (Raptor) were initially developed by our allies and chosen by the U.S. military to supplement our arsenal. The AGM-142 is a medium range air-to-ground missile launched from the venerable B-52H. A smaller version is being jointly developed with Turkey and Israel that can be launched from fighter aircraft. This is a very good, workable example of our government buying into technology. It not only ensures us the best of weapons and technology but also ensures that competition remains a strong incentive in the marketplace.

Recent conflicts in the Gulf, in Bosnia and in Afghanistan have shown the worth of our missiles, smart weapons and boots on the ground. In an unpublished letter to *The New York Times* a few days after the September 11, 2001 attacks in New York and Washington, I was able to state confidently that our tactics, weapons and commitment would prevail in the operations that would unfold in response to that cowardly attack (see #6 in Documents).

Smart weapons purchased by smart buyers are essential ingredients of victory, not only in the war on terrorism, but on every battlefield where we could possibly be engaged. Development

SMART WEAPONS NEED SMART BUYERS

of these weapons, large or small, cannot stop and start according to whim. In order to hold the tactical advantage we must continue to support and fund their development in good economic times and in bad. This also means supporting innovative manufacturing techniques, co-operating with our allies in developing mutual systems and "buying into" the data created in the research and development programs of our own defense contractors.

JR. MIDSHIPMAN

REAR ADMIRAL

GOING TO SEA

THE USS IOWA BB 61   Source: DoD Archives Public Domain

WITH GENERAL JAMES L. JONES COMMANDANT OF THE MARINE CORPS AT THE COMMISSIONING CEREMONY OF THE USS SHOUP

RUN SILENT RUN DEEP

COMMISSIONING THE USS BOONE

SMART WEAPONS CIRCA 1918 - THE KETTERING "BUG"
Photo Courtesy of the USAF Museum, Dayton Ohio

CHAPTER 7

HONOR ABIDES HERE
Gettysburg and Other Lessons in Homeland Defense

In May of 2001 I had an opportunity to walk the battlefields of Gettysburg as a guest of retired Army General Dennis Reimer. He was the Army Chief of Staff from 1995 to 1999, capping a distinguished career that spanned well over thirty years. I know him to be one of that rare breed of scholar-soldiers, those who have learned the lessons of history *and* applied them in modern warfare. He entertained and educated us all with his gift for bringing alive the drama and tragedy of that defining moment of our country, the Civil War. From what I have seen, Denny could have a second career as the history chair in any leading university!

I sit on the board of several defense companies and it happened that in early 2000 Dennis Reimer was asked to fill a vacant seat on the board of one of those firms. While attending board meetings near Gettysburg, Denny arranged a tour of the battlefield for some of the board members and our wives. Prior to our visit he sent along copies of the Civil War classic *The Killer Angels* (Michael Shaara, *The Killer Angels*. Ballantine Books Reprint, 1993) for us to brush up on our knowledge of this defining moment in American history. Walking the hallowed ground of Gettysburg and pondering the heroics of Joshua Chamberlain and all the other brave citizen soldiers who fought there brought home to me an absolute truth: no people fight so bravely as those who are fighting in their own fields and forests, in defense of their own homes and family.*

With homeland defense now a paramount issue in Congress and in the minds of the American people, images of a new reality are being raised, not of soldiers fighting a uniformed enemy in the orchards and wheat fields of Pennsylvania, but rather images of emergency response specialists in decontamination clothing, cleaning up a post office after a terrorist attack.

No nation can ever meet every possible threat of terrorist attack but we must be able to absorb and contain simultaneous attacks on our cities and critical facilities. These attacks will likely be in the nature of chemical, biological or nuclear attacks, the last-mentioned more likely to be a "dirty" weapon rather than a nuclear blast. It is therefore imperative that America develop the technology and purchase the right equipment to deal with a broad array of threats, technology and equipment that will allow us to get in there as quickly as possible and save as many lives as possible. We cannot do that on any large scale at this time.

This equipment should have the same priority as our pressing needs within the military itself, such as modernizing our fleets and aging aircraft or rebuilding our cruise missile stocks. We cannot wait until the day we need the equipment to put in our order with the manufacturer.

We would do well to remember that in the 1995 Oklahoma City bombing, the FBI could not talk to the fire or police service. This would also appear to have been a problem in the rescue

---

*Joshua Chamberlain was a teacher turned warrior and commanded the 20th Maine Volunteer Infantry Regiment at Gettysburg in the pivotal battle of Little Round Top on the second day. He fought in twenty-four major battles in his career and was wounded six times. He was given the honor of receiving the surrender of Lee's army at Appomattox.

Chamberlain went on to serve four terms as Governor of Maine. He died in 1914 and was buried at Bowdoin College where he had been educated and later served as president. He has been called by biographers "the very model of the citizen soldier" and the words "Honor Abides Here" are engraved on his tombstone.

efforts at the World Trade Center on September 11, 2001. A terrorist attack in our homeland, and the emergency radios won't talk to each other? This must *never* be allowed to happen again. The fix isn't that hard, but it is resource limited. We could do a lot with a little if we thought our way through this. The military has always dealt with a high-low mix on modernization and has some lessons learned to offer on how to fix it. We need to do a front-end analysis on who needs to talk to whom and then figure out how to make that happen with the limited resources we have. What we can do is to establish *national* standards and tie federal grants to compliance with those standards. This will in time assure communications compatibility.

The first Homeland Security Director, Governor Tom Ridge, has shown great initiative in getting control of the situation and establishing a baseline for action. He is getting the necessary funding to start fulfilling his mandate and therein lies the heart of the problem regarding the procurement process. The money that will now flow in to Homeland Security must be very tightly controlled and budgeted to ensure that first, we understand the problem, and second, we understand what constitutes a response, *before* we start throwing too much money at it. Harmonization is the key.

There are two areas of concern in defending American soil, preventative actions and reactive actions. The first concern is before the fact and the second concern, we hope, will never happen.

It would seem then that funding better security at borders and also funding intelligence gathering on the ground are good solutions to the preventative need. Reactive is after the fact and in that case we have to be able to get our people in and start saving lives. That is priority number one, but we cannot spend more precious lives to do so. We *must* provide for the safety of the people who go in on the ground after the fact. At the time of writing it is uncertain if this will include specialist units of the National Guard or the military under the command of the newly formed Northern Command (NORTHCOM) but it will certainly include the local first responders such as police and fire departments. These men and women are likely to be the first to arrive on the scene of any disaster and in emergency should fall under the jurisdiction of the Office of Homeland Security.*

When it came my turn to answer our nation's call to battle in Vietnam, I went into combat zones on more than one occasion with highly trained warfighters, entering enemy-held camps. Our troops were able to suppress the enemy due to four important factors: good training, good equipment, good co-ordination, and good execution. Having worked with soldiers of this caliber it is obvious to me that for future situations, both at home and abroad, we should expand our current units of highly mobile, anti-terrorism-trained soldiers. We must also expand their training to include working closely with local ERT (Emergency Response Team) units and to respond quickly to any city, facility, border crossing or port in America.

Some of the work that I did while in Vietnam involved the analysis of intelligence reports from throughout the Mekong Delta area, and formation of plans for interdiction based on that information. Intelligence gathering and analysis is agonizing and often dangerous work and the nature of the business makes it closely guarded stuff that does not lend itself to sharing. That

---

*There are over one million firefighters in the United States, of which approximately 750,000 are volunteers. Local police departments report an estimated 556,000 full-time employees of whom about 436,000 are sworn law enforcement personnel. Sheriff's offices report approximately 291,000 full-time employees, including approximately 186,000 sworn officers. There are more than 155,000 emergency medical technicians registered nationally. That is a combined total of over two million essential-service personnel we can draw on throughout the country. (The White House, Homeland Security Facts and Figures. http://www.whitehouse.gov/homeland/

was true in Vietnam among the various branches of Military Intelligence and was also true of the CIA and FBI and other agencies that did not share intelligence or assets concerning the September 11 attacks. It is only recently in our country's history that we have begun to understand fully the implications of not sharing intelligence amongst all branches. It is a delicate matter to balance the need for absolute secrecy and the need to share intelligence. Once two people (or two agencies) know something, the odds greatly increase that three will soon know. Ensuring they can and do communicate quickly and decisively with each other must be a priority in future.

Our business and community leaders must encourage young men and women to honor the call to duty and serve their country, and make it possible for them to do so, be it through President Bush's newly formed USA Freedom Corps or in the National Guard and the Reserve forces. These three groups will form a big part of our emergency response to homeland security and they must be trained and equipped to survive and carry out their jobs in contaminated areas.

The logistics and planning needed for this new task will be significant and they must be as encompassing as possible to prevent overspending and misappropriation. My past experience with procurement, and more lately in business, has given me a unique opportunity to observe the procurement process from both sides of the fence. During both careers, I have seen that when Congress agrees to allocate *any* urgent funds, be it for the military or for hurricane victims or other disaster relief, much of the money is spent in such a way that it does not improve the long-term situation. It serves rather as a stopgap action and is not a solution in itself.

Having a central procurement authority follows good business practice. I know that if planning and purchasing in the

Navy were left to each Commander at each Naval Base our Navy would be a *very* different Navy today! Leaving the details and execution of homeland security procurement to each state cannot be a cost-effective way to protect the nation.

A central procurement authority would ensure that emergency equipment and training for homeland defense teams in Area A are compatible and interchangeable with emergency equipment and training for teams in Area B. It would encourage and ensure competition and competitive pricing that is unavailable to small-volume buyers.

Savings gained by the Navy during my tenure as the Competition Advocate General of the Navy made by purchasing "off the shelf" were significant, and this is still a very desirable aspect of procurement purchasing today. Take a look at a single item like a handheld two-way radio. Emergency and Police services use literally tens of thousands of these items. They will be needed in bulk if we are to have an effective communications network for homeland security. If we buy these units ten at a time in each local authority we will pay near retail prices and it is unlikely they will be compatible with the radios of other Emergency Services personnel who may be flooding into the stricken area from every state in the union. If we buy 10,000 at once and an option on 10,000 more within four years we will have the *same radio* in all fifty state jurisdictions *and* we will get a better deal by being able to take advantage of simple market driven economics!

To illustrate this point is quite simple. During my time as Competition Advocate General of the Navy, I had opportunity to work with my counterparts in both the Army and the Air Force. I worked most closely with Army Brigadier General Charles R. Henry, by sending some of our Navy people over to show them the ropes. General Henry and his team had no trouble catching on to competitive contracting and neither did their suppliers. In July of 1986 the Army awarded Magnavox a contract for 4,692 handheld radios with a provision for four options to purchase up to an additional 17,773 radios at negotiated contract prices. The radio was NATO-compatible, allowing seamless communication with allied forces, had a longer battery life and was cheaper than other handheld units available at the time.

It drives home a point that business lessons, like history lessons, must be applied to ensure sound results. It is also true, both in the military and in the business sector, that in order to have the best product you must have the best people, a principle that can also be seen from our walk on the battlefield of Gettysburg.

With us during that walk was the Honorable Steven S. Honigman, the former General Counsel of the Navy. I remember our chatting for some time about the need for effective co-ordination to carry out a major battle or campaign. It was fortuitous for the Union Forces descending on Gettysburg that General George Meade, who was certainly the man for the job, had been appointed above senior officers to take command of the Army just four days before that decisive battle. President Lincoln saw in General Meade the right man to lead the Union Forces, and as Commander in Chief Lincoln risked considerable backlash to appoint him. His decision was the correct one and it was Meade's brilliance at co-coordinating the movements of the large Union Army that carried the day. Meade however was a warrior adept at the set piece battle and his weakness was a reluctance to pursue the enemy once initial victory was assured. The more flamboyant and aggressive U.S. Grant, fresh from his own victory at Vicksburg, was eventually given command of the entire Union Army as due reward for his relentless pursuit of the Confederates wherever he met them in battle. Relentless pursuit until absolute victory is a lesson we would do well to remember in our war on terrorism.

It was during the battle of Vicksburg that one of the most unusual procurements in the entire history of the U.S. military came about. It is described by U.S. Grant in his writings and highlights the predicament faced when procuring weapons "after the fact" and the ingenuity of the American war fighter. It should be remembered that this kind of last-minute emergency adaptation was possible in the days of low technology warfare but is not possible today.

"There were no mortars with the besiegers, except what the navy had in front of the city; but wooden ones were made by taking logs of the toughest wood that could be found, boring them out for six or twelve pound shells and binding them with strong iron bands. These answered as coehorns (mortars), and shells were successfully thrown from them into the trenches of the enemy." (Ulysses S. Grant, *Personal Memoirs of U. S. Grant.* C.L. Webster, New York, 1885)

CHAPTER 8

DAMN THE TORPEDOES
Congress, The Senate and Procurement

I am writing this chapter in the wonderful library of the Army and Navy Club in Washington D.C., a "haven of heroes" as it has been called by *The Washington Post*. Its membership has included pioneers of aviation and space flight, past Presidents of the U.S. and no small number of men and women who will never be known to their country by name, but whose service was no less.

Across the street from the Club is Farragut Square, named after Admiral David Glasgow Farragut, naval hero of the Civil War and first Admiral of the Navy. In a city of monuments, his statue in the square is remarkable in that it was cast from the bronze propeller of his ship, the *Hartford*, on which he uttered the famous and often misquoted words, "Damn the Torpedoes, go ahead, Jouett, full speed ahead." This was the first statue dedicated to a Civil War hero to be erected in Washington. (James P. Duffy, *Lincoln's Admiral: The Civil War Campaigns of David Farragut.* John Wiley & Sons, 1997.)

His life story is no less remarkable than the statue that represents it and begins with a strange twist of fate that led him to service on his first ship, the *Essex*. Farragut's father had saved the life of an old gentleman whom he found adrift in a boat while out fishing near their home on Lake Pontchartrain, Tennessee, while the future Admiral was still a boy of eight. The grateful son of the dying gentleman was Captain David Porter and he was much taken by the kindness shown to his father. To add tragedy to sorrow, Farragut's own mother died of yellow fever during this period and the two were buried on the same day. Captain Porter, a man of great honor, made an offer to George Farragut to adopt one of the children of the Farragut house.*

Young David's brother William was already a midshipman and David was eager to accept the offer. His father too was aware of the great opportunity this was for another of his sons and reluctantly bade him goodbye. Captain Porter had vowed to be both friend and guardian and Admiral Farragut himself wrote many years later, "I am happy to have it in my power to say, with feelings of the warmest gratitude, that he ever was to me all that he promised".

An account of Farragut's life was published in 1892 by respected author and historian Alfred Thayer Mahan. Mahan's greatest contribution to historical literature is said to be his demonstration of the determining force that maritime strength has exercised upon the course of general history. Mahan describes the building of the *Essex*, which was Admiral (then midshipman) Farragut's first ship. The final sentence is the telling one: "The Essex was built in Salem, Massachusetts, by a subscription raised among the citizens … The building of the Essex was thus an effort of city pride and local patriotism; and the launch, which took place on the 30th of September, 1799, became an occasion of general rejoicing and holiday, witnessed by thousands of spectators and greeted by salutes from the battery and shipping. The new frigate

---

*Porter served in the Barbary wars [1801-1805] and other expeditions and battles, including three years as the commander in chief of the Mexican navy. In the Barbary wars he was taken prisoner in the capture of the USS *Philadelphia*. These battles are immortalized in the "Marines' Hymn." The hymn takes considerable artistic license, as any good Marine can tell you, in that "From the Halls of Montezuma to the Shores of Tripoli" has got it backwards. The actions on the shores of Tripoli came well before the capture of Mexico City and the Castle of Chapultepec, or the "Halls of Montezuma" as they are known to the Marines.

THE ARMAMENT TIDE

measured 850 tons, and cost, independent of guns and stores, somewhat over $75,000... Notwithstanding the zeal and emulation aroused by the appeal to Salem municipal pride, and notwithstanding the comparative rapidity with which ships could then be built, the Essex in her day illustrated the folly of deferring preparation until hostilities are at hand." (Alfred T. Mahan, *Admiral Farragut.* New York: D. Appleton, 1892)

It is interesting to note that at the time Farragut was first getting his feet wet on the *Essex*, the Congress of the United States was taking its first steps to streamline and modernize the procurement methods of the fledgling U.S. Navy, a process that continues to this day.

Two documents that are part of our collective history illustrate what I mean quite well, and can be found in the documents section at the end of the book. The first is an excerpt of a bill from Congress (Bill #47) in the year 1815 that is almost amusing in its simplicity (see #7 in Documents). The second is a Senate debate from 1984 (see #8 in Documents). In both cases the point makes itself. I apologize to the reader in advance for including such windy documents as these (although both are edited for brevity even at this length!) but I feel they give a very accurate picture of the general consensus throughout our history that someone *very* senior, with a very capable, dedicated team, who is directly answerable to Congress, needs to be in control of the entire contracting and procurement process.

Congress and the Senate are normally the first hurdles in the procurement process. Those bodies and the various appropriations committees are the last defense against waste and mismanagement of tax dollars. This is true not just for military procurement but for every single dollar spent by the Federal Government on our behalf, and rightly so.

Procurement should never be free of Congressional or Senatorial participation and oversight, and as mentioned previously, I had already served a year in the role of Competition Advocate General of the Navy when Congress mandated such a role for all the Services in 1984. My experience in promoting competition to Congress throughout my tenure included the preparation and presentation of four reports to that august body. That experience confirms for me that the unwieldy process of procurement should be rationalized *before* it becomes an issue in Congress or on the Senate floor.

Re-focusing our vast procurement process is an ongoing task, one that did not begin with my appointment to the job of Competition Advocate General in 1984 and certainly did not end with my retirement. It was a small step in the right direction and with a succession of highly capable leaders who followed me, willing and able to carry on the concepts and even expand on them, those small steps have created savings of literally tens of billions of dollars in the past two decades. Now would appear to be an opportune moment to make the next steps and rationalize the entire process across the services. Just as our suppliers are merging into more cohesive and more capable operating units, so must we.

Previous chapters described the long lead-in time of some major procurements and the immense costs involved in developing new weapons and bringing them into service. Senators and congressional representatives who provide oversight on the various committees dealing with procurement are often senior politicians who are aware of the ebb and flow of history and the military. If we tell them we need something or in some cases do not need something, we must be fully prepared to back up our requests to the dollar and to answer hard questions. While Competition Advocate General of the Navy, I preferred it this way. It assured that none of my superiors could ignore the reports

THE ARMAMENT TIDE

or the efforts of our team. Like any other major item in the overall budget, each item the military buys must be justified in Congress, not just in the first year that the system is tabled, but also for each subsequent year throughout the development and fielding period.

In the procurement crusade to get the right weapons at the right price, the Senate and Congress give us a voice for our programs that otherwise might not be heard so loudly. On the occasions that I made reports to Congress I took the approach of Admiral Farragut and carried on, full speed ahead. Often our course in competition advocacy and procurement was cluttered with minefields and we were able to navigate with some certainty through them, knowing that our position was one of strength and that most of our effort would get through.*

My last Report to Congress was in 1986. My retirement in early 1987 resulted in the task of reporting being given to Assistant Secretary of the Navy Ev Pyatt and the new Competition Advocate General of the Navy, Rear Admiral R.M. Moore, for fiscal year 1987. In the 1986 report I was able to write in the Executive Summary (see #9 in Documents):"Fiscal Year 1986 will be viewed as a lasting benchmark for Navy procurement. For the first time since World War II, over half our Navy procurement dollars were awarded on a competitive basis. Significant savings have been achieved in shipbuilding, aircraft, missiles, combat systems, spare parts and maintenance programs…the Navy and Marine Corps leadership can be justifiably proud of the success of their acquisition workforce in applying sound business judg-

---

*Another historical oddity of Farragut's famous comment regarding torpedoes is that he was actually referring to mines which were known as torpedoes in his day. When Farragut called out "Damn the torpedoes…" he meant "damn the mines" as we would understand it. "Torpedoes" were considered an ungentlemanly weapon in those days and are still an effective and controversial weapon in the littoral battle spaces today.

ment and effective management control to the Navy contracting process."

Of the ten identified Priority Objectives we had worked on, Competition Planning topped the list. We had met and exceeded the goals set by Congress and with a good vision of where we were going in the future, I was also able to issue the warning: "In today's environment, some companies can find it more attractive to pursue merger and take-over opportunities rather than invest in capital improvements."

When I made that statement in 1986 the world was on the verge of an information explosion that is changing the world still. In my last report to Congress I was also able to state: "We are in a race between the demand for information and the ability to store, retrieve, process and use it. The volume of data required to support the procurement process can deluge organizations."

Recently, retired Vice Admiral Arthur K. Cebrowski, who reports to the Secretary of Defense on Force Transformation, has been quoted as saying that "the most fundamental shift in the rules has been from the Industrial age to the information age…" (Tom Philpott, *Interview with Vice Admiral Arthur K. Cebrowski*. April 2002)

Now, almost twenty years after my warning to Congress, the nation has access to a generation of bright young men and women coming out of our universities who have grown up with computers and who can utilize them to their full potential. We should be actively recruiting, not just in the business world but also in the software world, for the graduates who have a demonstrated head for business and a desire to serve their country. It sounds cynical but is not meant to be so, when I say that it is much easier to attract the best and brightest when the country is threatened than it is in times of peace and prosperity

THE ARMAMENT TIDE

when they all head for Wall Street.

Being on record as a staunch supporter of Senate and Congressional oversight I must also go on record as saying that I am not oblivious to the power of lobbyists or the dreaded "pork barrel politics" that can creep in to the system. Mega-companies present a *very* powerful influence on our executive branch and it is often hard for the uninitiated to separate the wheat from the chaff despite the hot air blowing. Lobbying by big companies, pressure by consultants, etc., can be used to create a skewed version of actual results. It is my view that open and fair competition makes this less likely to happen, and careful planning in the contracting phase makes it less expensive when it does.

History shows that both Congress and the military have, in the past, adopted and adapted to a policy of change in order to meet new challenges or opportunities. The wholesale change in military strategy and warfighting technology being experienced in the early years of the twenty-first century gives impetus for new looks at old problems, just as it did 200 years ago.

CHAPTER 9

THE ENDLESS CRUSADE FOR SENSIBLE PROCUREMENT
Today's Blunders are Tomorrow's Bills.

"Transformation" is the new order of business for the military under the Bush administration. Transformation is simple to understand but complex to initiate and as with any major corporation undergoing a major re-structuring, it cannot be done overnight or without a certain degree of upheaval. Under the direction of Secretary of Defense Donald Rumsfeld, the purpose of transformation is to prepare our military for new roles and new missions as they are likely to evolve across a broad spectrum. The threat-based planning of past generations (still current thinking for some decision-makers) is, "First, show me the threat in order to convince me to fund you because if you cannot demonstrate a threat, then it must equate that you have no need of a capability to defend with." This is a bit like the ostrich sticking its head into the sand. In the transformed military, the ability to respond quickly and decisively to a broad range of threats will replace the set piece battle strategy that we became accustomed to in the twentieth century. This transformation *must* include the procurement and contracting process to ensure due diligence to future generations of warfighters.

On September 23, 1999, while campaigning for President of the United States, then Governor George W. Bush outlined his thoughts and his vision for the military in a speech to cadets at The Citadel, the historic war college in Charleston, South Carolina. It was clear from the speech that a shake up was on its way.

THE ARMAMENT TIDE

"I will begin creating the military of the next century. Our military is without peer, but it is not without problems.... A volunteer military has only two paths. It can lower its standards to fill its ranks. Or it can inspire the best and brightest to join and stay.... My second goal is to build America's defenses on the troubled frontiers of technology and terror.... The protection of America itself will assume a high priority in a new century. Once a strategic afterthought, homeland defense has become an urgent duty.... Yet today our military is still organized more for Cold War threats than for the challenges of a new century – for industrial age operations, rather than for information age battles. There is almost no relationship between our budget priorities and a strategic vision. The last seven years have been wasted in inertia and idle talk. Now we must shape the future with new concepts, new strategies, new resolve.... As president, I will begin an immediate, comprehensive review of our military – the structure of its forces, the state of its strategy, and the priorities of its procurement – conducted by a leadership team under the Secretary of Defense. I will give the Secretary a broad mandate – to challenge the status quo and envision a new architecture of American defense for decades to come. We will modernize some existing weapons and equipment, necessary for current tasks.... On land, our heavy forces must be lighter. Our light forces must be more lethal. All must be easier to deploy.... I will encourage a culture of command where change is welcomed and rewarded, not dreaded. I will ensure that visionary leaders who take risks are recognized and promoted.... I will expect the military's budget priorities to match our strategic vision – not the particular visions of the services, but a joint vision for change.... I intend to force new thinking and hard choices.... I will not command the new military we create. That will be left to a president who comes after me. The results of our effort will not be seen for many years. The outcome of great battles is often determined by decisions on funding and technology made decades before, in the quiet days of peace...." (George W. Bush, *A Period Of Consequences.* Speech at The Citadel, September 23, 1999)

The changes promised by President George W. Bush have been hastened by world events and are now being seen in practice. Unfortunately, as a result of the current transformation effort to adapt the military to a broader range of scenarios, the military were caught out in the process of developing a weapons system it no longer needs or feels it can no longer afford with the limited procurement dollars available. This is not new to procurement. An interesting example of transformation, system development and project cancellation can be seen today at the United States Air Force Museum in Dayton, Ohio.

The display tells the story of an ingenious inventor in Dayton at the close of World War I. His name was Charles F. Kettering and he was employed by Dayton-Wright Aircraft Company which was a small cog in the enormous defense industry supporting the Great War. By 1918 the country's defense manufacturers were in high gear and, spurred on by the government's cry for more and better weapons and delivery systems, Kettering invented an unmanned aerial vehicle (UAV) that could take off from a dolly mounted on a track. With a wingspan of fourteen feet and just over twelve feet long, it was powered by an inexpensive forty-horsepower gasoline engine, had a range of seventy-five miles and was controlled by a system of internal pre-set vacuum, pneumatic and electrical controls. After a predetermined length of time a control closed an electrical circuit, which shut off the engine. The wings were then released, causing the "Bug" as it was called, to plunge to earth where its 180 pounds of explosive detonated on impact. Although initial testing was successful, slightly fewer than fifty were built and eventually the project was cancelled due to a lack of enthusiasm in Congress or the military to fund further development. A replica of the "Bug" was built by volunteer staff and can be seen at the Museum, which is located at Wright Patterson AFB in Dayton, Ohio. (see photo in Appendix A)

The past dozen years have seen the cancellation of at least four major programs no longer seen as crucial to our evolving needs. The announcement in May, 2002, by Secretary of Defense Rumsfeld of the cancellation of the Crusader artillery system (see Chapter 12) is a case in point, as is the DD 21 destroyer program and the Seawolf submarine. Making the point from an historic perspective is the A-12 deep-strike attack aircraft that never found a home once the Iron Curtain fell. In each case the decision to end the program was made at the highest levels but the intricacies of each cancellation were unique.

The common theme of these programs was to save our limited procurement dollars. Another common theme was that they were *very* expensive to terminate. In most cases they were programs that in the eyes of the incoming Secretary of Defense and his advisors were not consistent with the evolving role of the military. Each change of Administration brings with it new people, new priorities and new challenges. Often a good portion of the uniformed leadership has wanted to see a program completed and the system delivered regardless of the changing circumstance but it is up to the political appointees to make that decision and to take the subsequent heat.

It is important to note that the cancellation of a program does not mean that the role goes unfulfilled. There are alternatives to the A-12 just as there are alternatives to the Crusader and Seawolf. There are alternative artillery applications that can utilize some of the $2 billion already spent on Crusader technology and

THE ENDLESS CRUSADE FOR SENSIBLE PROCUREMENT

certainly there are many other uses for the additional $9 billion already earmarked for Crusader, an artillery gun that is seen by the Secretary as being unsuitable in the broad sense of our needs.* Much of it, however, will be money lost.

This is also true of the DD 21, although the program is evolving more than it is disappearing (see Chapter 10).

In any case, those decisions are made by the Secretary and our focus here is procurement and not defense politics. We need only concern ourselves with the mechanics of contracting and procurement, not the strategy behind it. For procurement officials it is a case of "Theirs not to make reply, Theirs not to reason why." (Alfred, Lord Tennyson, *The Charge of the Light Brigade*. 1864).

This is where the Government has miscued in the way it has cancelled these programs. While a case can be made for dropping a program that is already under contract, it is still a costly thing to do. Penalties are a normal part of contracting and absolutely fair. Many of the contractors have had to make huge investments that they have anticipated amortizing over the life of a major system and then have suddenly found themselves with nothing to build. How much they are paid in compensation is a fair question. How the Government gets out of the program with the least damage to both parties is the biggest concern once the decision is made. That is the role of the contracting office and something that must be addressed very clearly. Contracts begin at the bargaining table and should end there if necessary. Certainly the most expensive way of getting out of a contract is to have the Secretary of Defense announce it unilaterally at a press conference!

---

*In Afghanistan to date there has been no deployment of an artillery unit. All fire support has come from aircraft or weapons carried in by the troops. This does not mean we will not need artillery in other battles we will face in the future. It is certain that we will. Massed artillery is still a necessary part of any major land battle.

THE ARMAMENT TIDE

The cancellation of any project should be done by the contracting officer responsible, by the Service involved, and with the company affected also taking a role. Those advocating the cancellation must be prepared with hard data before going in front of any Senate or House committee. Supporters of the Crusader program on Capitol Hill have referred to the cancellation of that program with disdain. An especially colorful quote from Senator James Inhofe (R), in whose district the Crusader was to be built, is a good example of the reaction to an ill-prepared recommendation: "There has been no credible analysis or testimony against fielding Crusader. The Defense Department has proposed several alternatives it says will improve the accuracy, lethality and deployability of our military. The problem is that most of what they propose only exists on a PowerPoint slide." (James Inhofe, Letter to *USA Today*, May 15, 2002)

Failing to involve the Service affected by the cancellation in the early stages can also lead to conflict, which, if it spills on to the Senate floor, can cause unnecessary embarrassment to the Government and the Services. Army Chief of Staff General Eric K. Shinseki, speaking before the Senate Armed Services Committee (05/16/02), informed them that in his opinion killing the Crusader program would create a "window of risk" for U.S. troops in the field "until we find a replacement system." Secretary of Defense Donald Rumsfeld in his remarks afterwards was diplomatic in his response, saying that "There are certainly honorable, knowledgeable Army Generals who will say yes – and I respect that. But there are also honorable Army Generals who will advise you that we should press ahead with new technologies. It has always been so." (Senate Armed Services Committee, *Hearing on Crusader*. Federal News Service, May 16, 2002)

Procurement is serious business and no place for the fainthearted. In a 1986 interview, fresh from releasing my report to

Congress and only a few months from retirement, I felt it necessary to speak about the difficulties faced in rearming America.

"The world's early navigators pointed out that two-thirds of the earth's surface is covered by water. They failed to tell us that in later years it would appear as though the other one-third of the globe is covered in congressional staff members, auditors, and any army of lobbyists with endless lists of programs demanding immediate attention.... You should recognize that the political and economic pressures on our nation's top decision makers are probably as heavy now as at any time in our peacetime history....The Navy and the Department of Defense are under intense scrutiny as the Congress and the Administration examine what we are doing, why we are doing it, and whether we should be doing it at all." (Stuart F. Platt, "Buying the Right Ships and Weapon Systems." *Program Manager*, November/December 1986).

To fully understand the implications of a poor exit strategy in regards to a contract one need look only as far as the A-12 aircraft. What began as a next generation replacement for the venerable A-6 Intruder deep-strike aircraft ended in acrimony, billions of dollars spent on a non-existent airplane, and further millions on legal fees. Who the villains are depends on which side you are on: either the consortium of McDonnell Douglas (now part of Boeing) and General Dynamics, or the Government as represented by then Secretary of Defense Dick Cheney, who cancelled the program on January 7, 1991. At the time it was the largest contract termination in the history of the Department of Defense. It is still before the courts today.

Response to the A-12 fiasco came from two camps. Proponents of the system called for the head of then Secretary of Defense Dick Cheney along with billions in additional payments and penalties. The Government side denounced the contractors as

THE ARMAMENT TIDE

incapable of delivering on the aircraft. The Project on Government Oversight, whose self-proclaimed mandate is "committed to exposing waste, fraud and corruption in … defense, energy & environment, contract oversight and open government," called the contractors to task.

The Project on Government Oversight stated that "McDonnell Douglas and General Dynamics first bungled a contract so badly that no weapons were produced, helping themselves to $3 Billion, and then mugged the public again for $1.5 Billion blaming it all on the government. These corporations ran over the public once, then stopped and reversed over it all over again. The defense procurement system clearly needs serious repair to prevent this happening in the future." (Project on Government Oversight; Concerns and Questions: *The A-12 Aircraft Financial Fiasco.* October 2, 1996. http://www.pogo.org/p/defense/do-961008-reform.html)

Secretary Cheney's rationale for canceling the A-12 was a series of reports that suggested the aircraft was 30% overweight, grossly over-budget and, as subsequently confirmed by the contractors, facing serious engineering problems. The government stand was to "terminate for default" the contract for development and initial production of the A-12 and demand repayment of amounts already paid to the contractor team.

Initially the contractors in their rebuttal won the day and had the order changed to terminate the contract "at the convenience of the government" and demanded payment and compensation totaling billions. The Government is not short of good lawyers, however. Steven Honigman, General Counsel of the Navy, and Gene Angrist, Deputy General Counsel, made an excellent move in selecting a young and upcoming navy lawyer, Harold Cohen, to work on the Navy's termination defense team for the A-12.

This is consistent with my own belief that you need to

identify and then assign the best young talent to solve problems so they can get hard experience early on in their careers. It is also my belief that they must be rewarded appropriately for their successes. Harold Cohen is now the Counsel for the Navy Space and Warfare Command.

The U.S. Court of Federal Claims, in the appellate phase of the trial, which is now in its second decade, recently reached a decision that the Navy Contracting Officer for the A-12, Rear Admiral William Morris (a friend and colleague from my days as Competition Advocate General), had a "legitimate basis" for terminating the contract. Earlier on, Judge Robert Hodges Jr, who ruled on the basis for terminating the contract, had found the government at fault for the *way* that Secretary Cheney had terminated that contract.

Speaking after the ruling, Richard L. Aboulafia, Senior Analyst and Director of Aviation Studies at the Teal Group, a defense consultancy firm, said it was a form of justice and pointed out that the $4.4 billion originally bid by General Dynamics and McDonnell Douglas for the A-12 was nearly $2 billion less than that offered by the competing bidders, Grumman Aerospace and LTV Aerospace and Defense. Furthermore, he added that this sort of bidding was a standard practice of the Cold War era.

I can confirm from my own experiences in contracting that companies would at times underbid on a contract knowing that, if there were cost overruns, they could get more money to cover themselves by using various loopholes and clauses in the contract. The pressure to land a big contract or simply the lust to win at any cost can lead to company executives making poor decisions in the contracting process. For government to tacitly support this sort of bidding strategy is not only bad for business but leaves the company vulnerable to playing catch up from Day One. It is shocking to me that we seem to be in a situation of forgetting our history and being doomed to repeat it.

In Washington, along with the Army and Navy Club, another of my favorite places to stay is the Willard Hotel. It is only a short walk from there over to the Ronald Reagan Building and International Trade Center, itself a wonderful facility designed by New York Architects Pei Cobb Freed and Partners. I recently made that short stroll in black tie, much to the amusement of the shorts-and-tee-shirt-attired tourists out for an evening stroll of their own, to attend the Eisenhower Award Dinner honoring the 2002 Award Honoree Condoleezza Rice. Dr. Rice is the National Security Advisor, a former Provost of Stanford University and Professor of Political Science. I have made reference elsewhere to our need to seek out the best and the brightest to serve their country and Dr. Rice is a shining example.

The Keynote Address was by General Richard B. Myers, Chairman of the Joint Chiefs of Staff. As you would expect, the guest list included most of Washington's movers and shakers. It was black tie all the way with politicians, journalists, a few celebrities and a good sprinkling of the Nation's defense establishment past and present resplendent in their uniforms and formal wear. What was most remarkable about the social aspect of the evening, other than the tinkling of jewelry and medals, was the buzz amongst the guests regarding the bold move by Secretary Rumsfeld to axe the Crusader in such an abrupt manner. Secretary Rumsfeld has been around Washington for as long as most senior senators and congressional representatives, having served as an administrative assistant to a congressman in the Eisenhower administration and then serving several terms himself as Congressman for Illinois. Clearly he has earned the respect and support of many power brokers in Washington over the years, and the conversation was lively and pointed.

It remains to be seen what the cost will be to cancel Crusader or when the final accounting will end. We can be sure from past experience that it will be very expensive and will serve as another repetition of the lesson that in procurement contracting, it is important not only to spell out how you buy, but also how you get out of buying. In a rapidly changing world we cannot afford commitment on a "lock, stock and barrel" basis. We must have an exit strategy and stick to it when a decision is made to terminate a contract. In the rush to execute the decision, the government is making the job of cleaning up the wreckage that much harder.

CHAPTER 10

CHARTING OUR COURSE TO THE FUTURE

> "The art of war is of vital importance to the State. It is a matter of life and death, a road either to safety or to ruin. Hence it is a subject of inquiry which can on no account be neglected." Sun Tzu, 400-320B.C.*

In our customary bow to history I have taken the quote above from the Chinese scholar and warrior Sun Tzu who wrote *The Art of War*, a seminal work that is still taught in war colleges and even a few progressive business schools today.

It is difficult to look back on our history and find a point in time where America faced as big a challenge in force planning as that being faced today. The scope for action is larger now than it has been in the past but easily identifiable enemies and obvious "lines in the sand" have, for the most part, disappeared.

While much of the communist world collapsed upon itself in the late 1980s and early 1990s, America and her industrial base grew stronger. From that uninterrupted industrial growth, we have reaped technological benefits that have given us the tools to meet and defeat our new enemies wherever it is they may choose to draw a new line in the sand. Having said that, it is worth noting that the Chinese have been studying the works of Sun Tzu for two millennia, the west for less than one hundred years. We would do well to remember this emerging giant in all our future considerations and defense planning (see Epilogue).

---

* *The Art of War.* Metro Books, 2002

CHARTING OUR COURSE TO THE FUTURE

With transformation, we are looking far into the future and to plan for that future is a formidable challenge. It is a challenge that is facing every corporation in the world today. How do we prepare for and adapt to a rapidly changing future?

Planning for the future and then budgeting for that plan is complex. To muddy the waters further, each Service in the military is supporting major procurement departments. The Navy, for example, currently supports separate procurement offices for Ships, Aircraft, Space, and Supply. This can result in the same work being repeated by each procurement branch of the Services for purchases of common items beyond "belt buckles and beans." Inefficiencies on this scale can cost substantial amounts of money that could be better spent elsewhere. The Department of the Navy has a budget that, if compared to revenues of the top companies in America, would place it at #8 on the Fortune 500 list.

In the many years I have observed the military budget work its way through Congress, there is more often than not considerable debate although there is also general consensus. Even for those persons outside the military or government loop and not privy to intelligence briefings, a thorough study of the world situation will usually confirm the thinking behind any given budget. This procedural ability of Government and military leaders to arrive at a consensus makes it possible to envision a single DoD procurement authority that enjoys the support of Congress and the different branches of the military. A single authority would allow us to present a united and strengthened hand at the bargaining table. My experience has shown me that this is a workable approach to a growing problem and something best done sooner rather than later, as the stakes are so high.

Major corporations planning their budgets have adapted quite readily to the speed at which modern business is conducted and are able to come to far-reaching decisions much quicker than government is able to. This is partly because corporations are not under the scrutiny of Congress and a seemingly endless stream of sub-committees. As indicated in previous chapters, I feel that the screening process of Congressional oversight helps the procurement process by providing for national debate on major purchases, but there is much to be done to rationalize this process and other methods by which we conduct our business.

To begin to understand the size and complexity of budgeting and procurement for the U.S. military one only need look as far as a few simple charts. The first chart is intended to illustrate the sheer size of the Department of Defense as compared to leading multinational corporations.

DEFENSE BUDGET AND FORTUNE 500 COMPARISON CHART	
2002 Defense Budget	$331.0 Billion (source: DoD)
2002 Wal-Mart Revenues	$219.8 Billion
	(Fortune 500 List)
2002 Exxon Mobil Revenues	$191.6 Billion
	(Fortune 500 List)
2002 General Motors Revenues	$177.3 Billion
	(Fortune 500 List)
2002 Ford Motor Revenues	$162.5 Billion
	(Fortune 500 List)

The dollar numbers attached to the Defense budget, although startling to the uninitiated, represent no more than 6% of our Gross Domestic Product (GDP) even during the most ambitious defense build-ups of the last half century! The following table shows a comparison of defense budgets and GDPs of individual countries. GDP represents the total market value of all the goods and services produced by a country during a specific fiscal year.

DEFENSE SPENDING AS A PERCENTAGE OF THE GROSS DOMESTIC PRODUCT (CIA World Fact Book, Washington Post; New York University.)

Russia	(1999)	5.00 %
China	(2000)	1.20 %
Saudi Arabia	(2000)	13.00 %
Germany	(1998)	1.50 %
Japan	(2001)	.96 %
USA	(2003)	3.50 %
USA	(1985)	6.00 %

Of less importance but equally interesting is the next set of figures. I should make clear beforehand that it is *not* my intention to suggest that we would have a better Department of Defense if we paid the Secretary more money! It does point out the relative imbalance in wages between Defense and the private sector. This imbalance will exclude a good many highly capable applicants from the selection process and from seeking employment in other areas within the Department of Defense. The same is true for key persons who have been offered, but are reluctant to fill, politically appointed positions.

SALARY OF TOP MANAGEMENT IN DEFENSE AND CORPORATE AMERICA

Donald Rumsfeld – Sec. of Def.	= $161,800 (2002 - DoD)
H. Lee Scott Jr. - Wal-Mart	= $11,613,000
	(2002 - Forbes Lists)
Lee R. Raymond - Exxon Mobil	= $32,653,000
	(2002 - Forbes Lists)
G. Richard Wagoner Jr. – G.M.	= $ 3,162,000
	(2002 - Forbes Lists)
William Clay Ford Jr. - Ford	= $ 3,700,000
	(2002 - Forbes Lists)

THE ARMAMENT TIDE

The imbalance in the pay scale of the civilian bureaucracy within the Department versus the private sector makes it very difficult for the DoD to attract and retain good, solid people, not only at the top levels but perhaps just as importantly, in mid-level management as well. This issue must be addressed and corrected by some "outside the box" thinking. Innovative solutions might take the form of annual pay increases, along with benefit and pension packages that reward long-term loyalty and performance, above and beyond the normal incentives now offered by such plans. Perhaps it is also time to re-examine income taxes paid by members of the Armed Forces. I would suggest that some further exemption or tax benefit for military and civilian personnel should be considered at the earliest opportunity.

This problem of recruitment and retention extends beyond the civilian sector of DoD to the military people on the ground. We are not looking for just any breathing body to fill a pair of boots. Almost everything needs a keyboard these days and high standards of recruitment will become even more pronounced in the future. Although I do not anticipate that a Marine's job will become physically easier in coming years, the skills and abilities of our young men and women will truly become a case of brain before brawn.

Recruiting this talent and then retaining it brings to mind the debate that occurred in 1973 when America began the transition to the all-volunteer force during the Nixon administration. Although his presidency was tarnished by the Watergate affair, Nixon was able to lead America out of the morass of Vietnam, was the first president to visit China, signed significant treaties with the Russians on missiles and trade, and was responsible for the founding of the Environmental Protection Agency. As well, his savvy in picking his cabinet raised to the national and international stage such players as Henry Kissinger, Caspar

Weinberger, Melvin Laird. and James Schlesinger. It is a shame that so much good work was tarnished by the folly of Watergate.

Prior to the all-volunteer force we could simply call up all the young talent we needed to fill the ranks, the downside being that we would only have them for a few years. With an all-volunteer force, the pool of talented recruits dwindles significantly but they are much more likely to stay longer in the service. We must rely on the deep sense of duty and patriotism of many Americans and what financial remuneration we are legislated. Having come up to flag officer rank myself, I can testify that none of us are in it for the money!

Sadly, I have seen literally dozens of fine people leave the service or government due to financial pressures. Bringing onboard our best and brightest directly from universities and military academies and then losing them to the private sector within the first decade of their careers, just when they are about to enter their most productive years, is a major impediment to long-term cohesion in the military. Retention is a major focus in the budgets for all branches of the military and while recent pay raises for the Armed Forces have done much to ease the situation, much remains to be done in this regard. For example, the Services lose pilots each year to the private sector, many pressured by the high tempo of operations requiring long stays away from home, most of them leaving for better-paying jobs in commercial aviation. We spend years and millions of dollars training our young men and women aviators to a level that is the envy of the private sector and then suffer by having too few resources available to retain them. This extends beyond pilots to other high-skill jobs within the forces, including sonar and radar technicians and a myriad of other skilled personnel. We *must* find the solutions to stem this bleeding from within.

THE ARMAMENT TIDE

From figures in the charts we have seen, one might be tempted to think that there is little reason to support defense spending on the scale we see occurring in the early part of the twenty-first century. The reasoning is understandable but unsound; we have an obligation to the defense of the free world. Simply put, "if not us, who?"

We also have an opportunity to gain further technological advantage over our potential foes during the process of transformation. Investing in technology, be it by Ford Motor Company, by Exxon Mobil, or by DoD, is a case of short-term pain, long-term gain. We must be in no doubt – and the evidence of our good work in Bosnia, in The Gulf War and in Afghanistan is there for all to see – that this transformation and rearmament effort is *essential* to preserve the freedoms we now have. In terms of GDP, the procurement budgets in the next few years will rival the budgets during the build up to the 600-ship Navy during 1983 to 1987. These budget increases were necessary then and are necessary now to ensure our capability, both in the short term and far into the future, to respond to a broad variety of threats.

The Navy budget for 2003 is a very good illustration of where transformation is taking us and in what areas the armament tide will be most strongly felt. The Navy has made a long-term commitment to transformation that is apparent in press briefing documents supporting the 2003 budget. These documents show that investing in technology and promoting better business practices are listed alongside readiness and personnel as building blocks of success for the future.

Force Sustainment is one area that is seen as a challenge to budget planning. While the Navy invests in transformational technologies some fleet attrition is going unanswered. The Navy would like to see 180 to 210 new aircraft per year but are slated for only eighty-three in FY 2003. It isn't until 2007 that the Navy goal of more than 180 planes per year will be attained, IF no one tampers in that area of procurement in the meanwhile. Ironically, even with an extra $992 million for procurement in Fiscal 2003 the Navy will lose seven aircraft from its current strength in that period even without allowance for aircraft lost to normal attrition!

On the up side, those aircraft in service will have more and better weapons to shoot with as there is a substantial munitions purchase slated to replace the +4000 PGMs (precision guided munitions) dropped in Operation Enduring Freedom.* Research and Development also wins a big increase as would be expected in this era of transformation, but even here there is give and take. The controversial but much anticipated V22 Osprey VTOL (Vertical Take Off and Landing) aircraft for the Marine Corps had $23 million cut from its budget, along with a huge cut of $444 million in Space and Technology. The Navy is re-focusing much of its R&D budget on the transformational technologies such as Joint Strike Fighter and the DD(X) ships of the future Navy. Also receiving a boost is unmanned technology such as the Global Hawk AUV.

The new legacy force we are investing in is showcased by the DD(X) ship program and the development of the "electric navy." Under Secretary of Defense for Acquisition, Technology and Logistics E. C. "Pete" Aldridge said of the program, "This program and its spiral development approach will be the model for Navy acquisition in the years to come. DD(X) is the Joint Strike Fighter equivalent for shipbuilding." (Department of

---

*PGMs are a family of weapons that include the Tomahawk Cruise Missile, the JSOW (joint stand-off weapon), the SLAM-ER (stand-off land attack missile, expanded response), the JASSM (joint air-to-surface stand-off missile), the JDAM (joint direct attack munition), and the LGB (laser guided bomb) and others.

Defense, News Release, *Navy Announces DD(X) Downselect Decision,* April 29, 2002)

Navy brass (or "Blue Suits" as they are called when differentiating them from the Civilian appointees) are also enthusiastic about the potential for the future fleet and while testifying on April 16, 2002 before a Senate Sea Power Subcommittee hearing, Chief of Naval Operations Admiral Vern Clark said the DD(X) program will serve as the vanguard for the surface fleet of the twenty-first century. "It isn't just about numbers. In addition to buying enough ships and aircraft, we must buy the right ships and aircraft, with the right capabilities for our future fleet." (Walter T. Ham, *CNO Says DD(X) Will Play Key Role in 21st Century.* Chief of Naval Operations: Public Affairs, April 16, 2002)

Another of our best and brightest, and one who is front and center in the transformation process, is Rear Admiral Jay M. Cohen, Chief of Naval Research. I have had opportunity to work with Jay Cohen and the Office of Naval Research and can confirm that he is the right man for the job. A sure sign of the faith the Navy has in his abilities was his command of the attack submarine USS *Hyman G. Rickover* (SSN 709) (see Chapter 5). It is a true hunter-killer of the deep and runs quiet and fast. Rickover has left such a powerful legacy that this submarine in particular holds a special place in the hearts of submariners. It is not my story to tell. I rather hope that Jay Cohen sits down and writes his own biography some day, but it illustrates my point to mention that I know of several of the operational assignments of the USS *Hyman G. Rickover.* The ship and crew performed under very difficult and dangerous circumstances, often operating in communist-controlled waters to gather information on Russian and Chinese Naval bases and operations. To this day those missions remain classified. The ship carried out more than a few cliff-hanger operations, and all I can say about them is that it was skill, God's will, and a good ship that returned them safely to home port. A truly capable man in a crucial role. Thankfully many of our good people like Jay Cohen still choose to forego the opportunities of civilian life to serve our country.

Admiral Cohen noted in a speech at MIT in 2001 that four key elements will be impacted in the transformation to the "electric" Navy: propulsion, auxiliary systems, antennas/flat-panel displays, and high-energy weapons. Some of these technologies will eventually make their way on to the future DD(X) family of ships. (Jay M. Cohen, *Developing Technologies for Continued Naval Superiority.* Speech to MIT Club of Washington, November 6, 2001)

It is important to clarify the term "electric." It should not be understood to mean that the new generation will run on fancy Duracells. Rather the DD(X) family will be powered by the Integrated Power System or IPS, which is the all-electric concept for future ships. The power is produced by the tried and true LM2500 series of gas turbine engines, acting as the prime mover for the generator rather than directly driving the propulsion system. Much of our current surface fleet utilizes the LM2500 as the prime mover for the propulsion system and requires additional engines to produce electricity for the ship and its systems. The transformation to the IPS streamlines this system. Funny how we come back to streamlining and centralizing the process, isn't it?

To further illustrate the synergy between Services that can be exploited in procurement, the AGS (advanced gun system) of the DD(X) concept will be capable of firing a land attack projectile with a range of 100 miles. Enhanced munitions are exactly the sort of transformational technology that Secretary of Defense Rumsfeld intends to develop with technology derived from the

THE ARMAMENT TIDE

axed Crusader artillery system. Ordnance and other munitions, as discussed in Chapter 6, are also the perfect opportunity for transformational procurement between services. Further down the pipeline will come Directed Energy Weapons and Pulse Weapons that will cripple an enemy's electronics and warfighting capability. These weapons alone make the Integrated Power System and the DD(X) essential, as current ships cannot support the electrical requirements of future electronics and weapons systems. This is not just a Navy issue. To develop these weapons and munitions for use across the services requires close co-ordination and long-term consensus.

I have been working on this chapter on Memorial Day weekend and perhaps it is fitting that the final paragraphs tell the story of one of the fine servicemen and women who, day in and day out, risk their lives for freedom. Transformation means nothing if we do not have men and women like Staff Sergeant Kevin Vance ready and able to carry the day.

Staff Sergeant Kevin Donell Vance joined the Air Force eleven days after graduating from High School. While serving as a Tactical Air Controller (TAC) with American Special Forces in Afghanistan, Sergeant Vance was team leader for a force that was sent in to rescue Navy Petty Officer First Class Neil Roberts, who had fallen from a helicopter in a previous assault. Onboard the rescue helicopter were thirteen Special Forces members and eight crew. They were immediately engaged in a firefight upon landing and Spc. Marc Anderson, Sgt. Brad Crose, and Pfc. Matt Commons were killed in the initial moments of battle, and three other team members were injured. Pinned down throughout the day and often having to call in close air support, the rescue team continued to take casualties including Senior Airman Jason D. Cunningham and Sgt. Philip J. Svitak, who both died of their wounds. When the extraction was finally made, America had suffered its worst day of the war and seven of America's sons were dead.

In the offhand manner of a veteran warrior, Sergeant Vance was modest about his own heroics and his words best describe the character of our fighting soldiers:

"It felt really good when I got back and my buddies said they were sitting around the radio listening. They were impressed that I never got emotional and was calm and professional the whole time. I tried to keep a monotone voice…I received a minor wound to my left shoulder. It is a shrapnel puncture wound. I didn't notice it until a day later when I woke up and my shoulder felt like someone punched me. I then looked at the T-shirt I was wearing that night and noticed it was blood-stained.

I went through so many different emotions, excited, mad, frustrated, sad, any other emotion you could possibly feel, you feel going through this whole thing. And I felt guilty if I felt anything was funny, like Sgt Walker's helmet with the holes in it (or) because we had lost members of our team.

Everyone out there just did his job. I just did my job, everything came natural and my training kicked in. There is nothing I could have changed about that day. Nothing we could have done different or better. I could not ask for a better group of guys to work with. I have trained for eight years to do this and now I had the chance to get to do my job _ that is reward enough. Everybody working together and the good Lord is what got us home." (Interview of SSgt Kevin Vance 25 March 2002 – Bagram, Afghanistan. Sworn statement. www.defensereview.com)

CHAPTER 11

GREASING THE WHEELS ON THE BANDWAGON OF CHANGE

> "Another damned, thick, square, book! Always scribble, scribble, scribble! Eh! Mr. Gibbon?" The Duke of Gloucester, upon receiving the second volume of *The Decline and Fall of the Roman Empire* from the author Edward Gibbon*

Events proceed at a quicker pace now than in Gibbon's day. Most of what I have tried to say to this stage, although timeless, is also time critical and most has been told in an effort to guide the reader through the fundamentals of the problem. Now we must have some clarity as to the direction we must take, as each year we delay in major procurement reform, hundreds of millions of dollars are at risk of being wasted or poorly spent. The military has made a long-term commitment to a policy of transformation and this must cover not only the tools we need and how we fight a war, it must also transform how we acquire the tools to do it with.

Winston Churchill, in writing his six volume set, The Second World War, almost matched Edward Gibbons in long-winded and detailed analysis of events. Churchill begins his opus in *The Gathering Storm* by telling of the years leading up to the Second World War. To make his point clear he declares at the outset a

---

*1781 (http://www.compapp.dcu.ie/~humphrys/FamTree/Gibbon/edward.historian.html)

GREASING THE WHEELS ON THE BANDWAGON OF CHANGE

"Theme of the Volume," which he states is "How the English-speaking peoples through their unwisdom, carelessness, and good nature allowed the wicked to rearm." Were I to have chosen a theme for this volume it might be, "Look to the future, with an eye to the past and do not allow the defense industrial complex to manage or dictate the procurement process."

In a note on supply organization dated June 6, 1936, while the wicked were rearming on the continent, Churchill foresaw a need for a unified procurement branch in the armed forces of Great Britain. Written in the context of the expected outbreak of war in Europe, much of it rings true today, although Churchill could never have imagined an enemy that would arm themselves with civilian airliners and use them as kamikaze weapons.

"The first step therefore is to separate the functions of strategic thought from those of material supply in peace and war, and form the organization to direct this later process. An harmonious arrangement would be four separate departments – Navy, Army, Air Force and Supply – with the coordinating minister at the summit of the four having the final voice upon priorities .... At the present time the three service authorities exercise separate command over their particular supply.... What is needed is to unify the supply command of the three services departments into an organism which also exercises command over the war expansion. The Admiralty would retain control over the construction of warships and certain special naval stores. This unification should comprise not only the function of supply but that of design. The service departments prescribe in general technical terms their needs in type, quality, and quantity, and the supply organization executes these in a manner best calculated to serve its customers." (Winston S. Churchhill, *The Second World War.* Bantam Books, 1979)

In pre-war Europe and, in fact, the entire free world, alarm

THE ARMAMENT TIDE

bells were ringing loudly when Churchill wrote his note. America seemed determined not to be dragged into another European war and there were few in Government supporting the rearming of America. Even so, our government was moving ahead quickly to strengthen our naval forces and our fledgling air arm. On the one hand, a head in the sand attitude existed and on the other were people who knew that sooner or later America would answer the call to arms, so the sooner we got armed, the better.

Throughout the 1930s, with the majority of Americans coping with the depression, faint war clouds were forming on the horizon. Cash for the military was exceedingly hard to come by and yet Congress and the military somehow found the money to embark on a plan to modernize our war fleet and begin construction of a fleet of aircraft carriers, starting with USS *Ranger* (CV 4) in 1931. Aircraft carriers were transformational technology at that time and some in the Navy were not convinced they should be putting any eggs at all into the "air wing" basket. To that end American shipyards began construction of a fleet of modern battleships in 1937, including the USS *North Carolina* (BB 55) laid down October 27, 1937, by New York Naval Shipyard, the USS *South Dakota* (BB 57) laid down July 5, 1939, at Camden, New Jersey, by the New York Shipbuilding Corp, and the USS *Massachusetts* (BB 59), laid down July 20, 1939, by Bethlehem Steel Co. in Quincy, Massachusetts.

It is only an historical aside, but very relevant to our discussion, that although these battleships were "new" constructions, they actually comprised two separate and dated designs. These ships and their 16" guns were required at the earliest possible date. By dusting off old but available designs, they could be put to sea much sooner than the planned but not yet ready Iowa class which began with BB 61, the USS *Iowa*, laid down at New

GREASING THE WHEELS ON THE BANDWAGON OF CHANGE

York Navy Yard on June 27, 1940. A stopgap measure that worked out well for the war effort, the USS *Washington* (BB 56) and many of her older sister ships survived the war and there is plenty of steel on the ocean floor that did not. Certainly though, it was not an economic or procurement success story in the normal sense.

The second of the North Carolina class, the keel of the USS *Washington*, was laid on June 14, 1938, at the Philadelphia Navy Yard. After commissioning, sea trials and training she became Flagship to Rear Admiral John W. Wilcox, and, at the head of Task Force 39, was sent to reinforce the British Home Fleet at Scapa Flow in early 1942. Scapa Flow was home port to the major British fleet protecting Churchill's island fortress. Without the vital sea lanes to keep supplies flowing, all would be lost for Churchill, Europe, and the free world.

The USS *Washington* started her Atlantic crossing under a dark cloud. One day out from port, while making way in heavy seas, the "man overboard" alarm sounded and a quick muster of the crew revealed that Admiral Wilcox himself was missing. No one saw him go overboard. Perhaps while taking some fresh air on the deck he suffered a heart attack and was then washed overboard by the rough seas that occasionally swept the ship. Despite an extensive search by accompanying vessels and aircraft and a spotting of his lifeless body, they were unable to recover the Admiral. The USS *Washington*'s entry into the war then continued with another dramatic incident only weeks later. HMS *King George V* collided with a destroyer, the HMS *Punjabi*, cutting it in two directly in the path of the oncoming USS *Washington*. With no option but to pass over the wreckage of the sinking destroyer, the USS *Washington* was rocked by explosions as the HMS *Punjabi*'s depth charges, triggered by their automatic depth sensors, exploded. The USS *Washington* suffered no major damage but did sustain minor damage to fire control systems and radars. The minimal damage was testament to the good work done by the naval architects and shipyard workers!

More firepower was desperately needed in the Pacific and despite the horrific shipping losses to U-boats in the Atlantic, USS *Washington* was ordered to the New York Navy Yard in Brooklyn, New York, for a thorough overhaul. She then sailed for the Pacific in the fall of 1942 via the Panama Canal. She arrived on station in time to support the Solomon Campaign where she fought with distinction in the first engagement in the Pacific between battleships, as she went toe to toe with the Japanese battleship *Kirishima*. A description of that battle in *The Dictionary of American Naval Fighting Ships* reads: "In seven minutes, tracking by radar, *Washington* sent seventy-five rounds of 16-inch and 107 rounds of 5-inch at ranges from 8,400 to 12,650 yards, scoring at least nine hits with her main battery and about forty with her 5-inchers, silencing the enemy battleship in short order." (James L. Mooney, ed., *The Dictionary of American Naval Fighting Ships.* Naval Historical Center, GPO, 1991)

She emerged relatively unscathed from that battle and continued to fight her way through the Pacific campaign. On February 1, 1944, while maneuvering under cover of darkness the USS *Washington* rammed the battleship USS *Indiana* (BB 58) as she cut across *Washington*'s bow to fuel escorting destroyers. Both battleships had to put in to port for repairs. After reinforcing the extensively damaged bow at Majuro Island, USS *Washington* departed for the Hawaiian Islands where a temporary bow was fitted at the Pearl Harbor Navy Yard. The USS *Washington* then continued on to the Puget Sound Navy Yard in Bremerton, Washington, where she received permanent repairs and an overhaul. She was quickly back in the battle, however, and under

GREASING THE WHEELS ON THE BANDWAGON OF CHANGE

the flag of Admiral Raymond A. Spruance (see Chapter 4), commanding the 6th Fleet, and engaged in protecting the all-important aircraft carriers in what was to become known as the Battle of the Philippine Sea. The USS *Washington*, six other battleships, four heavy cruisers, and fourteen destroyers deployed to cover the aircraft carriers, and on June 19th the ships came under attack from Japanese carrier-based and land-based planes. This lopsided battle became known as the "Marianas Turkey Shoot," and seriously depleted the Japanese naval air arm. In four massive raids, the Japanese launched 373 planes and lost 243. They also lost a further fifty bombers based on the island of Guam.

In one of the epic battles of American history, during February of 1945, the USS *Washington*'s heavy guns fired 16-inch shells shoreward in support of the landings on Iwo Jima. By spring she was off Okinawa, shelling positions there in support of the efforts to cut off the Japanese on Iwo Jima. With the Naval war in the Pacific well in hand the USS *Washington* was ordered to Puget Sound Navy Yard on June 28 and did not return to the Pacific war. She was held in reserve for many years and was finally struck from the Navy list on June 1, 1960.

NORTH CAROLINA CLASS BATTLESHIP (USS WASHINGTON BB 56)

Date Launched:	June 1, 1940
Displacement:	35,000 tons
Length:	729'
Beam:	108'
Draft:	38'
Speed:	27 knots
Complement:	1,880
Armament:	Nine 16" guns; twenty 5" guns; sixteen 1.1" machine guns

THE ARMAMENT TIDE

SOUTH DAKOTA CLASS BATTLESHIP (USS SOUTH DAKOTA BB 57)

Date Launched:	June 7, 1941
Displacement:	35,000 tons
Length:	680'
Beam:	108' 2"
Draft:	36' 4"
Speed:	27.8 knots
Complement:	2,364
Armament:	Nine 16" guns; sixteen 5" guns; 68 40mm., 76 20mm

IOWA CLASS BATTLESHIP (USS IOWA BB 61)

Date Launched:	August 27, 1942
Displacement:	45,000 tons
Length:	887'3"
Beam:	108'2"
Draft:	37'9"
Speed:	33 knots
Complement:	2,800
Armament:	Nine 16" guns; twenty 5" guns; various other armament

Now, just as in Churchill's time, the overall problem of acquisition is simply that there are too many people, going to the same place, on different roads. The apparent solution, which I support, is that we need to construct an acquisition and procurement superhighway, a route to the warfighter for both buyers and sellers that is direct, quick, and has easy to navigate on and off ramps.

Beginning with our efforts in the mid 1980s, great strides have been made in acquisition reform but even these sometimes bold efforts have failed to keep pace on the road to the future.

GREASING THE WHEELS ON THE BANDWAGON OF CHANGE

Now is the time to build our superhighway. To continue on the present road while our vehicle continues to build speed is an accident waiting to happen. In the last two decades we have seen companies in not only the defense industry but most sectors of business merging into stronger, larger entities that are formidable business adversaries and also formidable business partners. They bring an amazing wealth of talent and ability to the drawing table and an equal wealth to the bargaining table.

Currently Defense acquisition falls under the general umbrella of the Under Secretary for Defense (Acquisition and Technology) {USD (AT)}, and as mentioned previously is further subdivided into each branch of the armed services, with each branch being responsible for its own procurement needs. Common items for all branches are purchased by the Defense Logistics Agency (DLA) on behalf of the three services in a separate process. DLA is also answerable to the USD (AT). The success of the DLA in feeding, clothing and equipping 1.5 million men and women is a story of co-operation and co-ordination in business that should not go unnoticed here.

A recurring problem in all branches of Government is the relatively short-term nature of political appointees. An incoming Secretary or Under Secretary, for example, will usually have the first months of his term taken up by confirmation hearings. Further months are needed to be brought up to speed and then comes a period of two years in which he can play an effective role, followed by a final year in which he is a lame duck if it appears that the incumbent president will not be re-elected. Few serve as lengthy a term in office as Robert McNamara, who holds the record for longest serving Secretary (seven years, from January 21, 1961 to February 29, 1968), or as often as Donald Rumsfeld, the current Secretary of Defense, who has served twice (from November 20, 1975 to January 20, 1977, and again from January

THE ARMAMENT TIDE

20, 2001 to the present). Elliot L. Richardson seems to have had the briefest tenure in office, a short three months between January 30, 1973 and May 24, 1973.

I recently attended a World Affairs Council presentation in Seattle that featured Former Secretary of Defense Robert McNamara as the speaker. He served as Secretary of Defense under both Presidents Kennedy and Johnson, and prior to that was the president of Ford Motor Company. During his tenure as Secretary he had battled with Rickover over the propulsion systems for the new aircraft carriers then planned. Rickover was adamant on two points: that they should all be nuclear-powered, and that the power should come from a two-reactor plant. Not only was McNamara against nuclear power, he did not support the research that Rickover needed to produce a two-reactor plant for the Nimitz class. McNamara eventually forced a compromise and so our force got one less nuclear carrier than Rickover wanted and was "stuck," as Rickover saw it, with the conventionally powered USS *John F. Kennedy* (CV 67), which entered service in 1968 and is still serving proudly today (see Chapter 2). I chatted with McNamara for a few moments before his talk and, after explaining who I was and my association with both Naval Reactors branch and later on with contracting of Nuclear powered Aircraft Carriers, I gently chided him, suggesting the two-reactor carriers of the Nimitz class had worked out well for the nation. He politely told me that he had been opposed to nuclear power at the time he was Defense Secretary, but that that was a long time ago. He spoke in a very polished manner and it was only in the Q&A period following that I could see that, like Rickover in his later years at Naval Reactors, McNamara had reached a stage in his life when he was on point with the topics at hand but had lost touch with the main themes.

The whole transformation process for procurement should

GREASING THE WHEELS ON THE BANDWAGON OF CHANGE

be seen as nothing more complicated than any other large corporation revamping its buying department. Automakers, for example, have unified many of the component parts for their product lines and they are sourced from the same suppliers and parts manufacturers. It simply does not make economic sense to design or purchase separately the standard components for their product lines.

From a buyer's perspective, there is a risk that these mergers of independently successful product lines into a huge parent company will reduce the product lines deemed less important to the parent corporation's overall strategy. In other words, one or all of the merged companies and their products may suffer to benefit the parent company. Employees can be cherry-picked for other divisions, profits diverted rather than re-invested, and well-laid plans disrupted.

In late 1981 I assisted in the creation of a plan put forward to the Secretary of the Navy by the Navy Reactors branch to solidify the position and authority of the Director for Naval Nuclear Propulsion. Nuclear power was (and still is) a critical component to our country's defense and at that time was focused around the ever-present Admiral Rickover, whose ghost, I am sure, still wanders the corridors of power today, be those corridors in the engine spaces aboard our submarines or in the vast acreage of the Pentagon. Rickover was nearing the end of his many years of distinguished service, but in order to effectively oversee this critical aspect of our military and to ensure long term efficiency and stability of the command, a formal declaration of the status quo from the powers that be was needed to protect the office during and after any transitional period.

The creation of that plan in late 1981 led to Executive Order 12344 of February 1, 1982, signed by President Reagan, which reads in part:

"By the authority vested in me as President and as Commander in Chief of the Armed Forces of the United States of America, with recognition of the crucial importance to national security of the Naval Nuclear Propulsion Program, and for the purpose of preserving the basic structure, policies, and practices developed for this Program in the past and assuring that the Program will continue to function with excellence, it is hereby ordered as follows:...The director shall be qualified by reason of technical background and experience ... the director may be either a civilian or an officer ... active or retired (sec.2).... The director shall be appointed to a term of eight years (sec.3).... An officer of the United States Navy appointed as director shall be nominated for the grade of Admiral. A civilian serving as director shall be compensated at a rate to be specified at the time of appointment (sec.4).... Within the Department of the Navy, the Secretary of the Navy shall assign to the director responsibility to supervise all technical aspects of the Navy's nuclear propulsion work, including: (a) research, development, design, procurement, specification, construction, inspection, installation, certification, testing, overhaul, refueling, operating practices and procedures, maintenance, supply support, and ultimate disposition, of naval nuclear propulsion plants, including components thereof, and any special maintenance and service facilities related thereto (sec.7).... In addition to any other organizational assignments within the Department of the Navy, the director shall report directly to the Chief of Naval Operations. The director shall have direct access to the Secretary of the Navy and other senior officials in the Department of the Navy concerning naval nuclear propulsion matters, and to all other Government personnel who supervise, operate, or maintain naval nuclear propulsion plants and support facilities (sec.9)." (see #10 in Documents)

My recommendation for an acquisition superhighway would

follow a structural plan similar to that suggested by Churchill and discussed at the outset of this chapter, one that brings the separate efforts of the Services into a single operating unit. This central buying command should be headed by a new four-star berth of Chief of Defense Material Procurement. This Flag Officer would have a proven track record in acquisition. I would further suggest that this Defense Material Procurement Agency should be focused on those acquisitions with a face value of +$250 million to focus the program on big ticket, long-term systems that are so vital to our nation's defense. One caveat I would insert here is that I feel the only truly "unique" procurement is that of ships and submarines. As they are a key component of our military strategy we may consider a separate position within the new Defense Material Procurement Agency for a high-ranking Naval acquisition officer, answerable to the Chief of the DMPA, and whose sole duty is the acquisition of naval vessels. *All* other items of military acquisition – aircraft, missiles, electronics – are inter-service connected, and none are so specific to a single branch as to require a separate procurement process.

As with the case for the Director of Naval Nuclear Propulsion in 1982, my proposed Chief of Defense Material Procurement should have the support of an executive order establishing the Command as well as outlining the mandate for the agency. This is for the simple reason that there is nothing like an executive order to get people focused quickly and moving forward! Events and technology are moving much too fast in the twenty-first century to stick to our guns literally or figuratively.

A six-year posting, rather than the eight years we recommended for the Director of Naval Nuclear Propulsion, would be sufficient to ensure continuity through changes of administration. The reasons that we felt it important back in 1982 to appoint a Director of Naval Nuclear Propulsion for a lengthy term were two-fold: first, to ensure continuity in programs, and second, to enhance the desirability of the post for aspirants to the position. This reasoning is also applicable to the proposed Defense Material Procurement Agency. A four-star, six-year berth to finish a career would be a highly sought-after prize for every ambitious officer.

I expect there will be considerable resistance to my proposal to centralize the acquisition command, given that resistance to change is an integral part of human nature. My response to those who stand in the way of serious reform is that the money at issue and the urgency of our times require them to stand aside: good business sense demands it. Right from his early days Rickover himself was a master at overwhelming any opposition to his programs, and it must be remembered that he served as an Admiral *longer* than I served in the Navy! Those who stood in the way of nuclear power were putting on the brakes right at the on-ramp to the future and Rickover could, on occasion, run right over them without even slowing down. Yet at the end of his career, ironically, many felt he was in fact the one driving too slowly, hunched over the wheel of an old car, right in the middle of the fast lane.

The risk that corporations, the government and even the arts face, whether selecting a CEO, a SECDEF, or a Leading Man, is that the very people who have the talent and who can provide strong leadership often have strong personalities. Some entrenched leaders can make it very difficult to instill a culture of transformation as was the case with McNamara and Sir A.K. Wilson (see Chapter 4). As further example, my wife and I are great fans and supporters of the ballet, and we can recall the sad spectacle of Rudolph Nureyev in the last years that he danced. It was obvious that he lacked the vigor, enthusiasm and freshness of those waiting in the wings.

GREASING THE WHEELS ON THE BANDWAGON OF CHANGE

"The only constant is change" is a phrase that has been applied to many situations and perhaps it is also fitting here. We have seen how men with vision and courage have had great effect on the outcome of world events, how long-term preparation is vital to long-term security and how transformation itself must be ongoing, inclusive, and ever adapting.

In the fifteen years that I have been retired from the military I have pursued, fairly successfully according to my investment banker, a career in business. I have come to learn that a capital strength of America is the quality of our universities and schools that teach business. At the executive level in America there are some truly brilliant minds. Innovation and vision are key elements to any successful business and America continues to prosper as a result of that free thinking. The experience I have gained from the two distinct vantage points of my naval career and my business career has shown me time and again that the only way for the U.S. military to deal with big business is to behave like big business.

CHAPTER 12

MANAGING WAR AND PEACE IN THE 21ST CENTURY

> "…common sense calls for a procurement process that is prompt, equitable, and administered with a firm hand that allows for good judgment." (Stuart F. Platt, "Thoughts for the Bush Administration." Proceedings Naval Review, 1989)

Our country is governed from Washington and the landmarks of the legislative process are familiar to most Americans. The Capitol Building, the Library of Congress, and the Supreme Court are well used, both by tourists who visit them and by our government. A good deal of the legwork of government is done however in the time-honored tradition of horse-trading over dinner and drinks. An establishment much frequented by the power brokers of our nation is the Old Ebbitt Grill, just across the street from the Treasury Building. Described as a "virtual annex" to the White House, you may find yourself sitting next to a big league journalist from CNN* or across from such White House heavy hitters as press secretary Ari Fleischer. In years gone by you may have been sitting down to lunch or dinner with Presidents. Grant, Cleveland, Harding, and Theodore Roosevelt are all known to have dined there during their tenures

---

*While having lunch at the Old Ebbitt I had half an eye out for CNN reporter Barbara Starr to say hello. Many years ago, before she became a TV personality, Barbara was a staff writer for Defense News and had written a flattering piece on my work as the Competition Advocate General. I haven't seen her in many years and I enjoyed working with her, she is an astute journalist.

in the White House.

Washington regulars will tell you that June is no time to be in Washington, D.C. The city can be hot and muggy in the summer and so it was during my short stop there recently to research current procurement in the U.S. military. While walking from the hotel to the Army and Navy Club library, I took the opportunity to stop at the Old Ebbitt Grill for brunch; I can recommend the crab cakes and poached eggs on English muffins. Being a busy place they serve up plenty of strong coffee and the *Washington Post* to keep you from noticing the wait for your food. In this day of fast food restaurants and sanitized menus, we perhaps overlook the opportunity that old-fashioned service provides for catching up on the latest gossip. If you were privy to every intrigue and deal made or broken there you would have an interesting book indeed!

While awaiting my crab cakes and eggs, an article in the *Washington Post* caught my eye. The ongoing saga of the procurement story of 2002, the Crusader artillery piece (see also Chapter 9) was back in the news and by all accounts will likely be playing out for a few more years yet. The Army is well represented, perhaps to the point of being over represented, on inter-service logistics and they had legitimately pressed hard for the gun as a key piece of their battle strategy as they understood it at the time. Had the Army purchased *and* sold the Crusader better to Congress and the Secretary of Defense, we may have been able to find a satisfactory solution.

In this case I might be convinced to side with my Army brethren who supported the continuation of the program. We still face a significant threat from countries such as China, on which I will have more to say in my conclusions and summary. North Korea and Iraq, along with many countries sitting on the fence, also have large, capable armies and healthy defense budgets. It is certain there will come a day when we will be fighting large, set-piece battles and will place high value on a big gun that can slug it out with enemy artillery and still bring decisive firepower to bear in close support of the infantry. It is also certain we will not always have the advantage of aerial supremacy and thus will not always be able to provide close air support of our ground troops such as we have enjoyed in the last few conflicts. Adverse terrain, bad weather, distances to target and many other negative factors can negate the smart bombs and delivery systems we have been utilizing so effectively in recent battles. Our troops in the field cannot wait for clear weather; they must live and fight in the adverse terrain and their lives often depend on immediate response.

I am sure the troops in Afghanistan would have been grateful for a few Crusaders had they been available and I am sure they would have been put to good use! I feel that for what it will cost us to get out of the Crusader program, we could have negotiated a limited production instead and then simply stored the bulk of them against the day when we will require heavy artillery support.

In other procurement news making the rounds in Washington at the time, a Navy Memorandum of Understanding (MOU) had raised a few eyebrows. The MOU is an agreement to terms and conditions for transferring the construction of four LPD 17 class amphibious assault ships from Bath Iron Works, a General Dynamics shipyard, to the Ingalls and Avondale shipyards owned by Northrop Grumman Shipbuilding Services (NGSS) in exchange for construction contracts of four additional DDG 51-Class destroyers to Bath Iron Works. Under the Memorandum of Understanding signed by the Navy, the four DDG projects that were to have been built at Ingalls will be transferred to Bath Iron Works. The assault ships, including LPD 19, in the initial stages of construction at Bath Iron Works, will be transferred to Northrop Grumman. Despite the Navy's anticipation of significant net cost savings on these programs the main benefit appears to be LPD 17 program stability and cost savings by centralizing production at one shipbuilder, NGSS, and providing for improved workload stability at General Dynamics Bath Iron Works, which will build additional ships of the DDG class in its new, more efficient, land-level facility. In other words, two conglomerates have gotten together in a deal to iron out their contractual and workplace-related problems by re-writing the original contracts.

Most of corporate America can only dream of such a convenient arrangement. The fact that this program change has been approved at the highest levels of Defense is an indication that we are already feeling the pressure from the mega-corporations to tilt the playing field. Any possible short-term gain will put us at risk of losing what little diversification remains in our shipbuilding industry. By establishing a lead yard technical capability and concentrating it at certain yards we are further reducing our industrial base.

To be fair, we must recognize and credit the fact that these companies are a national asset and as such, we must support them when and where we can within the boundaries of good business practice. In the flood tides defense suppliers are swept along in the rush and the passage is fraught with risk: in the ebb tides they can be caught up in a struggle to even stay afloat. Defense is a volatile business in more ways than one.

Admiral Skip Bowman, Director of Naval Nuclear Propulsion, spoke in April 2002 to the men and women who manufacture our submarines. In his talk he illustrated quite well my comments above. "When people talk about America's military juggernaut, they might think about submarines and ships, but

I think about you. You're the people and companies that built those submarines and ships despite thin profits, a declining stock market, and decreasing defense budgets. On the barest of margins, your sense of purpose has kept you a part of our team". (F. L. Bowman, ADM. "Remarks at Corporate Benefactors Day." *The Submarine Review*, April, 2002)

Bowman also went on to show that the current procurement plan for the submarine force is unworkable if we hope to achieve true savings in our program. They are still facing the same problem we had addressed in the 1980s, namely, that in order to take advantage of market economies, we must find a way to award multiple ship contracts.

"Multi-ship acquisitions would provide significant savings compared to one per year. Coupled with multi-year and Economic Ordering Quantity leverage, real savings are evident. So we should use all means that any good businessman would propose - innovative contracting schemes to maximize economic order quantities and multi-year contracts - but we are not. We talk about acquisition reform, but we are not executing that reform. As a result, we're passing up huge savings, conservatively estimated at $100m per boat. We must ensure legislators and executive decision makers understand the pain that comes from putting off smart decisions on procurement strategy until tomorrow" (F. L. Bowman, ADM. "Remarks at Corporate Benefactors Day." *The Submarine Review*, April, 2002)

While in Washington and between sessions at the Army and Navy Club Library and other business, I had the pleasure of a meeting with an old protégé of mine in the person of Rear Admiral Robert E. Cowley III, the current Competition Advocate General of the Navy. I was a bit taken aback when I arrived for my appointment with the Admiral. As I walked down the corridor to his offices, I was confronted by yet another row of portraits, and while this is nothing unusual when walking the many corridors of power in D.C., what caused my surprise was to find my own picture among them! It is with no small pleasure that I discovered it there along with other members of the "Navy Business Hall of Fame." Perhaps they dusted it off for the day, knowing I was coming!

Bob Cowley had worked within my realm of control at Naval Sea Systems Command (NAVSEA) and later had additional good experience as Assistant Commander for Contracts at Naval Air Systems Command (NAVAIR). The intent of my Washington visit was to gather information rather than try to sway incumbents. So I did not discuss with Admiral Cowley or others my ideas for a new Defense Material Procurement Agency (DMPA), but by listening carefully, I found we were in agreement on a host of procurement issues. The Joint Strike Fighter (JSF), for example, was a program that both of us agreed was a good deal all round and could serve as a model for procurement methods. In a similar vein both the Admiral and I felt that Boeing, despite losing out in the JSF bid, would do its utmost to enhance the role and increase the capability of the F/A-18 Super Hornet in the meanwhile. This harks back to my warnings that the major defense contractors can bring formidable forces to bear in order to protect their own programs. To paraphrase Macarthur's famous quote to Congress, "Old Soldiers never die, they just fade away", I would use "old programs never die, they just fade away, very, very, very, slowly."*

In many ways the U.S. Air Force is leading the way in pro-

---

*Macarthur addressed a joint session of Congress on April 20, 1951 after he was relieved from duty as the Commander in the Korean conflict. President Truman fired him, fearing his over-zealous prosecution of the war could lead to hostilities with China and perhaps even Russia. Although many attribute the saying to him, Macarthur himself points out in the speech that the line is actually from an old barracks ballad.

curement today. Dr. Marvin Sambur, Assistant Secretary of the Air Force for Acquisition, has given the nod to a plan that enshrines two priorities for the Air Force Acquisition program. The priorities are Speed and Credibility, the cornerstones of the Air Force's "Agile Acquisition" effort which began in late 2001. In a press release dated June 11, 2002, Sambur has publicly declared, "The two over-arching objectives of this policy are to shorten the acquisition cycle time and to gain credibility within and outside the acquisition community… every action and decision by individuals responsible for program execution must map directly to, and further these two primary objectives". General Lester Lyles, Commander of the Air Force Material Command backs up Assistant Secretary Sambur's comments by stating, "This is a huge step toward freeing our managers to manage… one size does not fit all." Sambur hits the nail on the head with his comment, "These programs are very complex and we have to stop trying to "eat the elephant" in one bite. If we work with our partners, the war fighters, testers, technologists, budgeters, and logisticians AND develop these systems in increments, we will break these complex programs into manageable bites. That will allow us to deliver capability more quickly and give us a much better chance of meeting our cost and schedule goals." (AFPN, "New acquisition policy stresses speed, credibility". AFMC News Service, June 7, 2002)

Sambur has it right but I am somewhat dumbstruck that it even needs to be said. He is either feeling the pressure of the big contractors and trying to inspire his troops, which I doubt, or he is trying to spur them along with a rallying cry to move quickly in order to insure that the Air Force receives a sufficient slice of the new procurement pie. Further, he should be looking beyond the Air Force. I can only begin to guess at the long term savings and opportunities possible in procurement were we to put together our best people under one roof. Conversely, I wonder at the risk to that same process should we fail to field our best team and should we fail to keep them all pulling in the same direction.

The corporate sector also has its strengths, and skilled program management is often one of them. It is not surprising that in the corporate sector, efficiency is the watchword and needless duplication is anathema to the ultimate result, the corporate bottom line. We would do well to remember those lessons and pay attention to that corporate dedication to the bottom line. Both factor largely in the following pages.

The objective of most corporations and the military is to be number one. The difference is that one seeks to be the number one supplier of widgets and one seeks to be the victor on the field of battle. Unquestionably a much nobler aim, but often thwarted by divisive tactics among military personnel as well as government officials. Personal glory is sometimes a roadblock to what should be the primary interests of the military and the nation. In World War II concessions were often made to some of the prima donnas so they could further their egos along with their legacy.

One of the reasons I have enjoyed reading Churchill throughout my life is his grasp of detail. He takes pains to enlighten the reader as to how things were accomplished in the pitch of battle. Tolstoy on the other hand, in *War and Peace*, is a master at describing why people do things. We follow armies into battle and we participate in their withdrawal from battle. We find all manner of descriptions of wars. What I find of interest for my purposes is that we never, in such a long work as *War and Peace*, learn how the logistics are accomplished or the differences in material support of the opponents.

Churchill does not simply give you a blow-by-blow account of a battle: he also outlines efforts that were set in place years

THE ARMAMENT TIDE

prior to the battle that had a major effect on the outcome. Churchill's own words give evidence that American war planners have long been able to think "outside the box":

"In the Pacific, the organization and production of the United States were in full stride, and had attained astonishing proportions. A single example may suffice to illustrate the size and success of the American effort. In the autumn of 1942, at the peak of the struggle for Guadalcanal, only three American aircraft carriers were afloat, a year later, there were fifty, by the end of the war, there were more than a hundred.* This achievement had been matched by an increase in aircraft production that was no less remarkable. The advance of these great forces was animated by an aggressive strategy and an elaborate, novel, and effective tactic." (Winston S. Churchill, *The Second World War*, Bantam Books, 1979)

The novelist can ignore the logistics and make his work no less interesting, but those who have to work on the question of "how do we get things done right?" do not have that liberty. Churchill did not have that liberty either in writing his history of WWII and we are the richer for it in both the literary and the military arts.

Defense contractors saw demand for their wares slide dramatically in the 1990s. Like any industry in contraction, competition became fierce. This was much to the benefit of the Pentagon, which had little need to put additional pressure on defense vendors. Now, with procurement expected to rise dramatically over the next several years, we need to change how we manage competition between the few defense vendors that have survived.

---

*The WW2 carriers were small by today's standard and *much* simpler than today's nuclear powered aircraft carriers. Still, such a massive effort was only possible because Roosevelt foresaw the need for aggressive rearmament and very early on encouraged industry to provide the shipbuilding capacity to meet the challenge.

Many of the factors that drove the defense vendors to compete vigorously have now been turned on their heads. As we have discussed at length, the procurement budget is growing as we reach into the first decade of the twenty-first century, not dwindling. However, other changes will have as much or even more of an impact.

The biggest change is the unexpected speed of the current flood tides of rearmament. We face new and immediate dangers that we are only partially prepared to face. This means that we need to take action quickly, building new capabilities and extending existing capabilities as they prove themselves against the new threats. This need for speed is a vendor's dream. Most people will have had the experience of telling a plumber or a mechanic, "I need it today." Most will also know that usually means a premium on the price you will pay. The defense industry is no different. For the next several years, our purchases will be made under the threat that delays could empower enemy success, either on the battlefield or in terrorist attacks. This takes away the buyer's option of delaying the awarding of contracts or other penalties if we start to see vendor performance slipping.

Another dramatic and significant change is the concentration of the vendor industry which has removed some of the options that had been previously available. We used the realistic threat of competition to great effect in the 1980s and with the merging of many of these companies, true competition is now much harder to achieve. In more and more cases, procurement officers will discover that there are only one or two surviving sources of a needed product or service, the rest having been gobbled up in the acquisition wave of the 1990s. As we have discussed earlier, there are ways of augmenting competition by helping to develop alternate sources of supply. However, these methods can often take substantial time, time that at this point in our history we happen to lack.

The surviving defense contractors have all been through major mergers; some have been through several (see chapter 1). The surviving companies are often coping with higher levels of debt than they would prefer. Now, these companies need to pay down that debt. This will be a major driver for those companies who are cash-poor, but contract-rich, to increase their profits.

As if high debt levels are not enough, the interest rates on that debt are set to rise. So too are the results expected by shareholders, providing a double whammy to the cost of capital, although not unique to the defense industry.

First, interest rates. The economic slowdown of 2001-2002 was fought off in part by an extended period of very low interest rates. Such low rates are never sustainable. Their rise directly adds to the interest rates paid by defense contractors: when old debt needs to be paid off, the companies must negotiate new debt at new, higher interest rates.

Second, shareholder expectations. In the late 1990s, financial markets (investors) experienced "irrational exuberance" in the words of Federal Reserve Bank Chairman Alan Greenspan. No more. Investors were then enamored of growth plans, e-anything, market share, and (apparently) aggressive accounting. Now, renewed investor caution is focusing on, of all things, actual bottom line profits. Gone are the days of investing for the future, now it is the era of "What have you done for me lately?" Consequently, we should expect the management teams of all companies, including defense vendors, to pursue every reasonable strategy to increase their short and medium term profits.

Though industries are different, the basic sources of profit are common. With higher debt loads, higher costs for debt, and increased profit expectations of shareholders, corporations are left with the choice of either increasing revenue or lowering costs.

The cost side will be hard for defense contractors to improve. Overhead costs have already been reduced through the shakeouts and mergers of the recent past. Production costs are somewhat more manageable, but not in the current environment. Production costs are easiest to reduce during times of stable, predictable production. Such conditions allow producers to scrutinize how they do things, working on the thousands of steps methodically. Nevertheless, in the current rearmament environment, production stability will be a casualty. Unless actively managed by vendors and procurement officials in a joint process, production will be erratic, as both react to the rapidly changing needs of the battlefield. Even if very well managed, production schedules and priorities will be far too dynamic for their cost declines to be a top driver of profit.

Though the vendors will make progress on managing costs to increase profit, the biggest opportunity for them is to increase revenue. Clearly, with more money being spent on defense procurement, getting as big a piece of that pie as possible will be the top corporate goal of all defense contractors. This will turn into two types of objective: to get the highest volume of work possible, and to achieve the highest possible prices. The drive for volume is good: it will motivate creativity, responsiveness, and all of the benefits of competition. The drive for volume is not good except for the vendors, their bankers and their shareholders.

As most of those managers know, raising prices is the single most powerful tool to boost profit. Unlike selling another dollar of widgets, with associated production costs, every dollar of higher price turns directly into a dollar of profit. In most businesses, the lack of pricing power has become one of the singular management issues of our era. Fragmented domestic competition and international competition via imports have consistently reduced corporate pricing power since the late 1980s. This fact is one of the key reasons that inflation was subdued during the late 1990s, with price deflation becoming accepted as normal for industries from manufacturing to technology.

That is not how it is going to work for defense contractors. Domestic fragmentation and foreign competition are not real inhibitors in this case, as the domestic industry has become quite concentrated. For many reasons, we simply will not be buying substantial amounts of military equipment from foreign suppliers. After ten years of slow procurement from vendors on the defensive, we are now facing the need to buy quickly from a concentrated industry facing high pressure to raise profits - and prices - any way they can.

I am going to go back once more to my favorite retreat in the Library at the Army and Navy Club. It is a wonderful place to think and to ponder both the past and the future. On the second floor of the grand ballroom, overlooking the dance floor, are many historical works of art. Among them is a splendid portrait of Captain Charles Sigsbee. If art is not of interest to you, the architectural splendor of the building alone is worth a visit. Judging by his portrait, Sigsbee was a man who must have been a striking figure on the bridge of his ship. A remarkable piece of Naval and American history is found in his story and that of the USS *Maine*.

She was commissioned as a Second Class Battleship and was launched in 1889 at the New York Navy Yard. She sailed into the port of Havana and into history on January 25, 1898. Spain was in a life and death struggle with her Cuban colony and the Atlantic Fleet was ordered there to show the flag and to protect American interests. On the evening of February 15 she was torn apart by a huge explosion and went quickly to the bottom of Havana Harbor. Of the 350 officers and crew aboard that fateful night, 252 were killed in the explosion and eight more died of injuries in the following days.

A court of inquiry immediately following the explosion could not find a cause for the disaster but the tide of public opinion in America was turned. The sinking of the USS *Maine* was used to add fuel to the fire by many influential U.S. newspapers at the time and the powerful Hearst chain of papers in particular trumpeted the loss loudly. "Remember the Maine" became the battle cry of the day and the U.S. declared war on Spain on April 22nd of that year.

The mystery explosion was re-investigated in 1910 under Congressional authority and the hulk was raised from the sea bed. The second board of inquiry determined that the damage to the bottom of the hull had been caused by an external explosion that then set off the forward magazines. However, there was much disagreement amongst the technical investigators at both inquiries, with most suggesting that the explosion was the result of coal dust combusting in the coal bunker which had, unfortunately, been installed adjacent to the reserve magazine for the 6" gun. In any case, nobody could determine who may have placed the explosives, if indeed any had been placed. Her main mast can now be seen overlooking the graves of some of the officers and crew at Arlington National Cemetery and her foremast is a memorial at the U.S. Naval Academy. Twenty-seven *Maine* crewmen are also buried in the beautiful setting of the Key West Cemetery in Florida, the last port of call for the USS *Maine* before the passage to Havana. Many years ago I was assigned to temporary duty at Key West and often passed by the cemetery on the drive to the base from BOQ (Bachelor Officer Quarters).

The final twist to the story came many years later when Admiral H.G. Rickover took it upon himself to reinvestigate the explosion

THE ARMAMENT TIDE

and sinking, using modern scientific knowledge and techniques. He published a book on his research titled *How the Battleship Maine was Destroyed*, published in 1976 (Francis Duncan & Hyman George Rickover, *How the Battleship Maine was Destroyed*. United State Naval Institute. Reprint April 1995) and in it he concludes that the most likely cause of the explosion was the coal and that the damage was inconsistent with an external explosion. It is remarkable that the Admiral found time in his busy schedule to take on the task and more remarkable for the fact that in many ways it defamed the institution of the Navy. He had to have known that he was seriously rocking the boat. He showed that technical failure rather than enemy action had destroyed the USS *Maine*.

I believe the Admiral knew what he would find beforehand and in his usual manner he conducted his investigation thoroughly and professionally. It was his own conviction (correctly held, in my opinion) that technical excellence and painstaking attention to detail and design must be placed on a pedestal. We get for our defense dollar exactly what we put into it. That is still true today.

EPILOGUE

VICTORIA PER PERSEVERANTIAM VENIT
Through Perseverance Comes Victory

"Fewer and better troops and simpler administration. Talks, speeches, articles and resolutions should all be concise and to the point. Meetings also should not go on too long." - (Mao Tse-tung, *Little Red Book,* 1949)

"Well, at least the Chairman and I could agree on some (administrative) points…" (Stuart Platt, 2002)

Formed in the melting pot of two centuries and more, America has welcomed the tired, the poor, and the huddled masses of the world (see #11 in Documents). Lady Liberty has greeted immigrants to our shores for well over a hundred years now, my own grandparents among them.* America gave them hard honest employment and an opportunity to seek success and they seized it with both hands.

Two cornerstones of our democracy, the Constitution and our Armed Forces, now protect 287,000,000 people and more, representing every corner of the globe and every religion. Our role on the international stage is part inheritance and part our choosing: in both cases we owe our allies and our nation our best efforts to ensure that America and the values we share with our global friends and neighbors are still around in another 200

---

*In the years between 1892 and 1954, approximately twelve million people entered the United States through the port of New York at Ellis Island.

THE ARMAMENT TIDE

years and much more.

The rearmament tide is beginning to flow at its strongest now and I am going to borrow from the melting pot principle for this last charge: everything gets thrown into the pot and you can take from it as much or as little as you like.

WAR ON TERRORISM

The War on Terrorism will continue to serve as the focus of national debate and of military action for the foreseeable future. Iraq, Iran, North Korea, the Philippines, Yemen, Somalia, Syria, and many other hotspots of terrorism must be dealt with. In the style of guerrilla warfare, the terrorists who survive will flee from one temporary sanctuary to another. Where we have co-operation we will work with local government to aid them in ridding their countries of terrorists. Where we do not have the support of the local government and they aid and abet terrorists or terrorist activities, we have rightfully put them on notice that the U.S. *will* take action to pre-empt any terrorism originating within their borders.

I concurred with the President when he stated, "Over time it's going to be important for nations to know they will be held accountable for inactivity, you're either with us or against us in the fight against terror" (George W. Bush, *President Welcomes President Chirac to White House.* Office of the Press Secretary, November 6, 2001) . My observations of the president's demeanor in handling himself in crisis make it clear to me that he has a personal resolve to see this through.

John Paul Jones is forever wrapped in the flag and the history of our country, and although a Scotsman, fought valiantly on the American side during the Revolutionary War. Jones raised the first Grand Union flag aboard an American warship, the *Alfred*. Historians dispute what flag was actually raised, but

EPILOGUE

the one most often seen in historical and romanticized accounts shows the flag bears the motto "Don't Tread Upon Me" and that reflects the attitude of our President and our Nation still.

If I were asked whether I was a hawk or a dove, I would say neither: I am a realist and a patriot. As much as we would wish that it were not so, our country will be more deeply engaged militarily in the next decade. Americans and American interests will be targeted throughout the world. The battle has already begun and now we must have the fortitude and the weapons to fight it to a successful conclusion.

Other areas we must pay very close attention to are the dangers presented by economic, psychological and cyber-terrorism. It is vital that we take steps to protect the infrastructure of America – dams, power stations, water supplies, and communications networks to name a vital few – but we must be very careful in the process not to create unnecessary fear in our citizens. On occasion, our media does us a disservice by giving undue attention to the smallest scrap of rumor and running with it, playing straight into the hands of the terrorists who count on *exactly* that sort of thing to create terror in the first place. Our leadership must show that it will be "business as usual" despite the risks that may be present. We will not tolerate having our cities and our citizens living "under the gun."

Cyber-terrorism is a newer and more difficult area to deal with. In an era when it takes seconds or less to communicate across the world and business and government rely more and more on internet-based communication and advanced computer systems to run their day to day affairs, it is easy to understand the chaos that could ensue were a cyber terrorist or foreign agency able to strike deeply via our computer and communications systems.

Terrorism is not just about killing people or subverting computer systems: the economic cost to our country to fight the

THE ARMAMENT TIDE

battle on terrorism is incalculable.* Without a sound economic base and a robust society we would not be able to wage this war at all.

This war will continue and other problems will evolve elsewhere in the coming decades and we can expect to put some heavy wear and tear on our ships and aircraft in the coming years. The Navy carried a heavy load in Operation Enduring Freedom and that with a fleet which, from a peak of 600 when I retired, now numbers just over 300 ships in 2002. Despite decades of effort by the Navy and some members of Congress who have been trying to convince the powers that be to fund and maintain a standing fleet of modern, large deck, aircraft carriers, we are in danger of finding ourselves under-strength and overstretched (see Chapter 2). Carriers have the singular ability to take the fight to the enemy anywhere in the world, at a time of our choosing. One reason alone should convince skeptics that we need to increase the number of large-deck aircraft carriers: where else can we expect to be able to conduct air operations from? Without forward airbases to stage from (which aircraft carriers provide), it would be impossible to obtain critical air superiority in a reasonable and sensible timeframe. We simply do not currently have the right to land in or even overfly many countries that might be located in potential conflict regions. Political considerations prevent us from using the land-based facilities of many of our "allies" in the Middle East. From the Red Sea, the Mediterranean and the Arabian Sea, our carriers can dominate the airspace there and engage our enemies any-

---

*Susan Page, writing in *USA Today* on May/02/2002 in the article "Sept. 11 attacks hurt some industries, helped others" stated: "The economic impact of the war on terrorism will depend on how long it lasts, how much it costs and whether it slows the trend toward globalization. If this war continues for years, as President Bush warns it will, analysts say it could have the most far-reaching effects on the U.S. economy of any event since World War II."

where in the region they may care to set up shop.

The overall Congressional support needed is now in place to carry forward our shipbuilding programs. New carrier constructions should be at the top of the list, funded in the next budgets, and the first keels laid as soon as possible. Even if we start fast, it would be *seven* years before the first vessel would see duty. Can we be certain that in seven years we will *not* need the additional strength? Carriers are arguably our most valuable asset in the War on Terrorism *and* in a conventional war, and I believe we are under-strength already.

To make a very simple comparison, conducting a war on terror calls for some of the same strategy as developing a good investment policy for a family's hard earned dollars:
- It is important to think long term.
- Diversify: do not put all your eggs in one basket.
- Do not let volatility of the moment affect the long term strategy.
- Carefully planned strategies tend to work slowly and unravel if hurried.
- Be patient, keep your focus on the end game: it is easy to get unrealistically excited by short run gains.

CHINA

China presents a wildcard in defense planning. I have had opportunity to travel widely in China and have visited most of their major shipyards. I was surprised and in awe of their approach to shipbuilding. The yards were for the most part self-contained and to my western eye looked like large prisons! I had a chance to look closely at the quality of the work in everything from hatch construction to cutting steel and altogether they do a good job. They have the workforce, the infrastructure, and the capability to build large ships in a hurry, and if you can build

THE ARMAMENT TIDE

big commercial ships, it is only a small leap to convert to the construction of large warships.

Large is one thing and quality another, however, and electronics in particular are a key area where China has fallen behind and we must be very careful in what technology we allow to be exported. The ability to produce warships that have the capability to sustain in battle and to prevail will take time, but the literature is out there for study. Recent failures in the Song class guided missile submarine development has left that program in limbo, but China is known to have aircraft carrier construction as a priority and has also gone ahead with the purchase of four Russian Kilo class submarines as well as placing orders for eight more. This threat from the sea *cannot* be overlooked or underplayed. Taiwan, Japan, and India in particular are vulnerable to a resurgent Chinese Navy.

China seems to be pursuing a two-pronged modernization strategy. The first prong seeks to enable them to take Taiwan by force of arms. The other prong seeks to enable them to prevent any intervention by the United States, which is committed to Taiwan's defense. Until now, the U.S. has been able to operate with relative impunity in the region but the risk to our fleet will go up considerably when the twelve Kilo class submarines are operational.

As well as having a thriving defense industry of its own, China was also the world's biggest importer of weapons in the first years of the new millennium, according to the Stockholm International Peace Research Institute (Stockholm International Peace Research Institute. http://www.sipri.se/)

While the true economic impact of China is just starting to be felt, the impact of the PLA (People's Liberation Army) has always been substantial. The PLA has played a key role both politically within China and beyond, and as a military power

EPILOGUE

both within China (against its own citizens) and beyond. Regional sensitivities in this area are a concern and there is more than one tinderbox awaiting a spark. Although large in numbers, the PLA has not posed an insurmountable threat to date. Thank goodness for the Himalayas, which (particularly in the twentieth century) have served to keep China out of India and beyond.

The modern PLA presents a danger to be sure. It has a capable nuclear force and in sheer numbers a strong military, but has not had the ability to move far beyond China's own borders. Internal strife has also served to keep China focused more inward than not. As China and the Chinese leadership come under international pressure to open their borders to reform and economic investment, a new reality will emerge: an economic giant, with a modern, well equipped military, an educated, global-looking leadership, and a huge question mark hanging over them. *How* will they integrate into the global neighborhood: by force of arms or by economic expansion?

I recently had opportunity to hear from two men who are well placed to anticipate and understand the future direction of the Chinese government. I met with Edward Nixon, brother of President Richard Nixon and an old China hand. He is CEO of Nixon World Enterprises, Inc., a firm that specializes in developing international trade and investment opportunities. Ed has made some thirty trips to China over the years as a consultant, and we discussed at length the implications of a resurgent Chinese military. He agreed that the PLA in particular is still a major player in Chinese political thinking and not to be underestimated.

Rep. Mark Steven Kirk (R-IL), himself an experienced combat aviator, stated in an article published in *Sea Power Magazine,* June 2002 "The third issue of the 21st Century is the rise of China. It is said that the 20th Century was domi-

THE ARMAMENT TIDE

nated by the emergence of the United States as the pre-eminent world power. The 21st Century will see the continued rise of China. What direction will it take in its foreign affairs, and how will the international system as a whole handle the growth of Chinese income and influence? …with the growth of its economy, there is a real possibility that China eventually will expand its military capabilities. This is a critical issue—not just for the United States." (The Navy League, "The Best Defense is a Devastating Offense." *Sea Power Magazine*, June 2002)

He is being realistic in expressing his concerns and I would add that we must also give consideration to China's ambitious space program. They have declared an interest in establishing a space station as well as a base on the moon and have timelines in place to accomplish those goals. This factor alone is enough to convince me that our nation must continue to develop a space-based missile defense system. The free world cannot risk being held hostage from space.

THE FLOOD TIDE OF REARMAMENT REPRISE

At the risk of flogging the horse at the finish, it is worth having another brief look at the principal points we have covered on procurement and procurement reform: after all this is my area of expertise! I will not try to cram in a lifetime of experience here, but will instead direct those wishing for more detail to Appendix B where the points below are covered in detail along with additional information and summary.

Competition amongst vendors for defense dollars is good for the military and good for America. A moderately fragmented vendor base is more competitive and delivers greater value for the buyers of any goods. The defense vendor market is more concentrated now than it was the last time (1980s) the U.S. embarked on substantial rearmament. To achieve good value for taxpayers' money, procurement should be managed in such a way as to increase competition.

During previous flood tides of rearmament, many of the capital ships and systems were bought under multi-year contracts, sourced from a single supplier. From a strictly business point of view, most CEOs would be alarmed by the implications to the shareholders of single-source supply to their company's production line. We were as well, and used basic management principles to fix the problems. One of our favored tactics was to encourage existing vendors to compete in new categories related to their areas of expertise, when we only had single-source supply.

1. Competition works. To achieve good value for taxpayers' dollars, manage procurement to increase competition. Resist the temptation for extra long term contract relations. They are comfortable and breed inefficiency.
2. Get the most out of vendors by requiring that procurement professionals have a deep understanding of the businesses of their vendors.
3. Defense is big business. React accordingly.
4. Major programs, especially technology development programs, take place over many political cycles. Manage such programs for stability, by retaining good people and adequately rewarding them.
5. Manage evolution of technology over multiple generations of products via long term focus and management accountability.
6. Use distinct contracting practices for distinct product types and phases of a product's life cycle. For example, use cost-based contracts for R&D to insure we then own the patent and data rights. Keep production under firm fixed contracts

THE ARMAMENT TIDE

to allow us to hold the contractor's feet over the fire.

ENDING WITH A BEGINNING

I am going to end this chapter and thus the book with the story of a new beginning. A singular example of every topic covered in this book is the USS *Shoup* (DDG 86). The USS *Shoup* is about procurement, good and bad: it is about rearming, it is about the War on Terrorism, it is about transformation, and it is all about the men and women of the American military, whatever their branch of service.

The USS *Shoup* is the latest of the Aegis class of ships to be commissioned (see Chapter 4) and that milestone took place in the shadow of the Seattle Space Needle, a target of a terrorist attack that was thwarted by some very good work by our intelligence services. For that reason alone, Pier 37 in Seattle was a most appropriate venue for the traditional, solemn and yet joyful occasion of the commissioning. This ship and her sisters will be at the forefront of the war on terrorism in the coming years: guarding our carrier fleets, interdicting suspect shipping on the high seas, launching her long-range missiles in support of Allied action and otherwise protecting and projecting the principles of democracy wherever she sails.

Her abilities as a warship could not have been imagined during the boyhood of her namesake, General David Monroe Shoup. General Shoup earned the nation's highest award, the Medal of Honor, while commanding the Second Marines, 2d Marine Division, at Betio during a bitter fight for control of Tarawa Atoll during WWII. The British presented him with their Distinguished Service Order for this action. One of the ships supporting the Second Marines at Betio was the USS *Sigsbee* (DD 502), named in honor of the Captain of the USS *Maine* (see chapter 12). Shoup went on to become Commandant of

EPILOGUE

the Marine Corps. His photograph, in full dress uniform, hangs in the officers' wardroom of the ship. As well as serving as the dining area for the officers, it is also the informal meeting room and the "living room" of this large extended family, and very different from the cramped and basic wardrooms found on most Navy ships of previous generations.

It should be understood that a ship's commissioning is different from the launching ceremony, which is more familiar to civilians with its traditional breaking of a bottle of champagne on the bow. The commissioning occurs not when the ship first hits the water, but when it has completed its trials and is akin to "handing over the keys." The Captain and crew are then considered ready to take the ship into active service. As part of the ceremony, three flags are raised simultaneously: the Stars and Stripes, the Naval Union Jack, and the Commissioning Pennant. Following these is the raising of the flag of the Senior Officer present, in this case the flag of General James L. Jones. It was always an honor for me when I was Senior Officer to see my own flag raised and I suspect it was no less so for General Jones. (see photos)

This time honored tradition of the flag raising gave a new twist to an old ceremony and in so doing aptly demonstrated the advent in the military of technology and transformation. When the command to raise the flag was given there was no hauling of halyards as the crew ran the flags up the mast: rather it was the simple push of a button and up they went!

In another old Navy tradition, the first crew of a ship has the honor of being referred to as "Plank Owners." In the Program for the ceremony are photos of each member of the ship's crew: representative of our melting pot of society, these men and women are a reflection of the modern Navy. The "Plank Owners" of the USS *Shoup* are first class sailors on a first class ship. As I

THE ARMAMENT TIDE

looked at each of the faces, I marveled at the future that is laid before them, taking a new ship into the fleet, to serve in defense of their country.

I spoke briefly at the ceremony with the "father" of the Aegis program, Wayne Meyer. He has an exceptional understanding of potential incoming threats to ships in a war zone and how to fight the ship against them in response. I know Wayne well from my years as the Competition Advocate General and I strolled over for a chat at the reception. He got in the opening gambit by greeting me with mock horror and a "Here comes the Competition Advocate General, I'm lucky we even got a ship!" I therefore reminded him that when Lehman and I had left for the private sector we had left him (Meyer) on watch and we had a 600-ship fleet! Wayne knows darn well if it were left to me they would have funds for Aegis and more. We both know that the likelihood of the USS *Shoup* being at the forefront of the battle in the coming years is very high. It was a sobering thought as we stood there among the eager young crew. They will be tested in battle and they know it. I was pleased to see Admiral Natter spending time with the crew and speaking personally with as many of them as he could.

The USS *Shoup* is also one of our key deterrents, should we ever be pressed by China to take stations in the Far East. I have personally worked in Taiwan and understand the risk to our Allies in the Far East. As part of a team headed by Earl Fowler I assisted in advising the Taiwanese Navy on radar systems for their ships and was instrumental in helping them select a phased array radar that employs similar technology to much of the Aegis electronic hardware. I know whereof I speak when I say there is not a ship afloat that can match the Aegis ships in battle. The USS *Shoup* is the most advanced of the Aegis class and has benefited from the latest upgrades slated for this "flight" of construction.

EPILOGUE

Our potential adversaries like China already know about the USS *Shoup*: they have had to take into account this ship and her sisters in every scenario they might consider outside their own borders since the keel was laid. From that day forward the USS *Shoup* has been paying her way.

A telling fact of the cost of freedom can also be found in the Official Program from the commissioning of the USS *Shoup*, which shows a copy of the Department of Defense form DD250, signed by the shipyard, the Navy inspectors, and Ship Captain. This is the form companies send on to the Naval Finance Centers to receive payment under their contracts (one thing can be said for the process, the government pays its bills on time!). The number in the amount box was $1,025,105,000. That is one billion, twenty five million, one hundred five thousand dollars, and no cents. I estimate the real number is higher because of change orders, support costs, spare parts and the myriad other expenditures that have to be made to bring a ship to life.

I have been to more than a few commissioning ceremonies and this one was notably different in the obvious and serious security presence. The U.S. Coastguard kept a vigilant eye on proceedings from the waterside, the National Guard, some with bomb sniffing dogs, as well as the ship's crew replete with battle jackets and a startling array of firepower including shotguns and automatic weapons, were on the dock. They had a stern demeanor and were obviously ready to use their weapons and that is as it should be, given today's circumstance. There was also a large assortment of city, county and state law enforcement officials, and the usual assortment of guys in bad suits with earpieces, keeping watch street-side. I was struck by the fact that without exception, everyone in the security detail had a weapon locked and loaded.

There must have been 2000 or more guests and everyone had

to pass under the gaze of literally dozens of eyes before getting anywhere near the ship. I did not see any exceptions and that included the obligatory electronic scanners at the various entries. Fortunately, it was a beautiful sunny day and everyone was happy to be out. The ship itself was a beautiful sight, bedecked from stem to stern with flags to mark the occasion. Knowing the key speaker was General James L. Jones,* the Commandant of the U.S. Marine Corps, the crowd likely figured (incorrectly) that they would not be kept waiting through interminable dull speeches and so the wait to pass through security was one of greeting old friends and catching up on the gossip. When the clan was finally gathered** Jones himself spoke like a Marine, short, sharp and on point, he read freely from what looked like hand-written notes and engaged the audience directly.

One lingering thought from that day sticks with me quite clearly. General Jones, when speaking to the gathered crowd for the commissioning, summed up the situation our nation faces when he told them, "If we want to be the home of the free we will have to be the land of the brave."

---

*General James L. Jones is the 32d Commandant of the Marine Corps. Jones is a highly decorated Vietnam veteran with a strong record of dedication and service to his country.

**Also present at the ceremony were the ex-Chief of the Joint Chiefs of Staff, General John M. Shalikashvili, as well as the Commander in Chief, Atlantic Fleet, Admiral Bob Natter. Bob's wife Claudia was co-sponsor of the Commissioning Ceremony.

GLOSSARY OF ACRONYMS

(acronyms for weapon systems are given in tables and charts in the text)

AGS ....Advanced gun system
CEO ....Chief Executive Officer
CinC ....Commander in Chief
CG.................Guided Missile Cruiser (ship)
CG(X) ....Future Guided Missile Cruiser (ship)
CNN ....Cable News Network
CNO ....Chief of Naval Operations
CVN ....Carrier Vessel Nuclear
CVNX ....Carrier Vessel Nuclear Experimental (ship)
DDG ....Guided Missile Destroyer (ship)
DD(X) ....Future Guided Missile Destroyer (ship)
DLA................Defense Logistics Agency
DoD ....Department of Defense
ERT ....Emergency Response Teams
G.E. ....General Electric
GAO ....General Accounting Office
GTGS ....Gas Turbine Generator Set
HMS ....His/Her Majesty's Ship
IPS ....Integrated Power System
JSF ....Joint Strike Fighter
LHA................Landing Helicopter, Amphibious (ship)
MOU..............Memorandum of Understanding

NAVAIR ....Naval Air Systems Command

NAVMAT ......Naval Material Command

NAVSEA ....Naval Sea Systems Command

NASA ....National Aeronautics & Space Administration

NGSS ....Northrop Grumman Shipbuilding Services

NR.................Naval Nuclear Propulsion Program

NORCOM ....Northern Command

PLA ....People's Liberation Army (China)

SAM ....Surface-to-air missile

SC ....Supply Corp

SecNav............Secretary of the Navy

SecDef ....Secretary of Defense

SHP ....Shaft Horsepower

SSN ....Ship, Submersible, Nuclear

SSBN ....Ship, Submersible, Ballistic, Nuclear

SUPSHIP........Supervisor of Shipbuilding and Repair

TRW ....Thompson Ramo Wooldridge (defense contractor)

USD (AT) ......Under Secretary for Defense (Acquisition and Technology)

UAV................Unmanned aerial vehicle

USAF..............United States Air Force

USN ....United States Navy

USS ....United States Ship

VLS ....Vertical Launch System

DOCUMENTS
Personal letters
Fitness report
1986 Report to Congress

1

THE WHITE HOUSE

WASHINGTON

December 24, 1986

Dear Admiral Platt:

Once in a while I receive word about a special individual whose example inspires others. That's exactly how I felt when I heard about your patriotism and service to our country.

Throughout your naval career you played a critical role in the preservation of America's security and freedom, as well as in the pursuit of world peace. As you retire after 31 years of service, I am proud to speak for all Americans in thanking you for your dedication to duty.

Nancy joins me in sending our congratulations and warm best wishes to you and your wife Melonee.

Sincerely,

Ronald Reagan

Rear Admiral Stuart F. Platt, USN
Supply Corps
Department of the Navy
Washington, D.C. 20360

2

THE ASSISTANT SECRETARY OF THE NAVY
(Shipbuilding and Logistics)
Washington D.C. 20350
9 June, 1983
Commodore Stuart F. Platt, SC, USN
Deputy Commander for Contracts
Naval Sea Systems Command
Department of the Navy
Washington D.C. 20362

Dear Stuart:

As I approach the end of my tenure as Assistant Secretary, I would be sorely remiss if I did not express my personal appreciation to you for your counsel and assistance in the execution of my duties.

I accepted the challenge of returning to Government service because I sincerely believe that our nation's security critically depends on the health and vitality of its maritime armed forces. It will always remain a source of pride for me that during my tenure in this office the serious decline in our strength was arrested and, working together, we have put the nation's sea services back on the road to achieving the maritime supremacy on which all of our nation's security interests revolve.

I especially appreciate your support, Stu, in rationalizing our new construction and overhaul contracting. Your contributions are appreciated within the Navy and in the private sector, as well. I am confident you will continue to render superlative service to your country in the years ahead.

Best of Fortunes.

Sincerely,

(signed) George A. Sawyer

THE ARMAMENT TIDE

3

Fitness Report: Commander S.F. Platt
Reporting Date: May 1974
Reporting Officer: H.G. Rickover

"As administrative contracting officer, Commander Platt directs the administration of Navy contracts totaling about $3.5 Billion with Ingalls Shipbuilding Division of Litton Systems at Pascagoula, Mississippi. The contracts involve construction and overhaul of major naval vessels including the construction, overhaul and refueling of Nuclear Attack Submarines. Because of his competence and proven ability, he was specifically selected for this important assignment, a particularly difficult one in view of the numerous claims being prosecuted against the government by Litton Systems under its shipbuilding contracts. During his first year as administrative contracting officer, Commander Platt has done an outstanding job of administering contracts under my technical cognizance. He has demonstrated professional skill, imagination and initiative in resolving several long-standing contractual issues, and in instituting programs to help foreclose additional claims against the government. He has made a concerted effort to close out old contracts so they could not be used as a basis for further claims against the government. To date nine contracts representing over $350 million have been closed, the first contracts to be closed since Litton acquired the Ingalls shipyard over 10 years ago; an additional $100 million of contracts are in the final stages of close out. Commander Platt is effective in identifying deficiencies in the contractual and financial aspects of the contractors operation. He meets regularly with the supervisor of shipbuilding and the president of Ingalls to resolve these issues and has made considerable progress in eliciting corrective action. For example, he recently successfully resolved a four year dispute over insurance recoveries, which resulted in $2.3 million reduction in Navy shipbuilding costs. Commander Platt is intelligent and carries out his responsibilities in a professional manner. He represents the Navy effectively in complex accounting and contractual disputes with senior company executives. In view of his outstanding performance I strongly recommend him for early promotion."

(signed) H.G. Rickover

DOCUMENTS

4

Department of the Navy
Headquarters Naval Material Command
Memorandum for: Commander, Naval Ship Systems Command
Commander, Naval Supply Systems Command
Subj: Assignment of outstanding Supply Corps officers to key contracting officer billets at major private shipyards.

1. In view of the complex contracting issues and the claims situation at our major private shipyards, you and I agreed that two outstanding Supply Corps officers with proven records and extensive experience in contracting and contract administration should be assigned to take charge of the contract functions at each of our major private shipyards – Newport News, Ingalls and Electric Boat. Based on advice from you and Admiral Rickover, I personally selected the following officers for these important billets:

CDR. S.F. Platt   SUPSHIP Pascagoula

LCDR J.C. Krummel

CDR B.M. Cole   SUPSHIP Newport News

LCDR D.E. Ledwig

LCDR J.R. Bartel

I understand they have already been assigned to these positions.

2. I consider these contracting officer positions at the shipyards are among the most important in the Supply Corps and would normally be filled by officers of the rank of Captain. Since these billets are vital to the Navy's shipbuilding program, the Commander, Naval Ship Systems Command should have a hand in the future selection of Supply Corps officers to fill these billets and their orders should provide for additional duty to the Commander, Naval Ship Systems Command. Appropriate arrangements should be worked out between the two commands.

3. These assignments may disrupt the normal career rotation of the officers involved. However, these assignments are being made at my specific request for the overall good of the Navy. This fact should be recognized by future selection boards in their deliberations on promotions and accelerated promotions. To make these positions attractive for officers in the future, special

consideration should be given to rewarding outstanding performance in these positions by early promotions.

4. Please ensure that the above information is properly documented and recognized so that future selection boards will fully appreciate the importance of these jobs to the Navy and the reasons why these individuals were selected to fill them.

5. A copy of this memorandum should be included in the service jacket of the above named officers so that the importance of their assignments is understood by members of the selection boards.

(signed) Admiral I.C. Kidd

SUPSHIP Groton

5

United States Atomic Energy Commission
Washington 20545
28 June 1969
Admiral B.H. Bieri, SC, USN
Commander, Naval Supply Systems Command
Department of the Navy
Washington D.C.

Dear Admiral Bieri:

I recently detached LCDR S.F. Platt, SC, USN, after 6 years in the Naval Reactors Program.

Platt did an outstanding job, both at our Naval Reactors field office at the Atomic Energy Commission's Bettis Atomic Power Laboratory in Pittsburgh, Pennsylvania, and here at Washington. He is one of your most intelligent and capable officers.

You should be proud to have an officer of his ability in the Supply Systems Command.

Sincerely,

(signed) H.G. Rickover

DOCUMENTS

6

Letters to the Editor
The New York Times
229 West 43rd Street
New York, NY 10036
Dear Editor:

As the former Competition Advocate General of the Navy during the Reagan administration, as a graduate of Stuyvesant High School in lower Manhattan and as an American father whose daughter walked out alive from World Trade Center Tower One, I would like to express to the people of New York my sympathy and my pride during this trying period we face in the aftermath of the atrocity inflicted upon us all on September 11.

As we move forward in our plans to bring the perpetrators to justice we would do well to take heed of Rudyard Kipling's words written for other soldiers so many years ago...

When you're wounded and left on Afghanistan's plains,

And the women come out to cut up what remains,

Jest roll to your rifle and blow out your brains

An' go to your Gawd like a soldier.

That is still the enemy we will face today, determined, fanatical and ruthless. We want to think long and think smart as we get in a fight on the ground with these people.

Having said that, I must take issue with the press pundits and others who promote the premise that our Armed Forces could not do better than the Russian Military or the poor British soldier in Kipling's poem. We are not sending into battle disillusioned young men who are poorly led and poorly motivated.

As our Navy Aircraft Carrier battle groups take station along the Tropic of Cancer in the Arabian Sea, we will have tactics and tools developed by Colin Powell, Jay Johnson, Denny Reimer and others during the Gulf War that have been practiced, honed and perfected continually since, We have new and upgraded weapons systems that have come on-stream since the early 90's, and we have another generation of great leaders in men like Admiral Vern Clark and General James Jones. We will not lose!

We will have the support of our many friends and allies around the world, who have now bravely stood to be counted as defenders of real freedom and justice. They are many and they are also strong. The United States of America will form a coalition, which will not flinch in the face of terrorism now or in the future.

Our priority as citizens of the United States is to support George W. Bush, our President, to continue to build a strong America for our children and to be patient for justice which may be a long in coming, but will sweetly come none the less.

Our military priority will be determined by the reasonable men and women who lead the Armed Forces, we will suffer casualties, we will see our sons and daughters committed to the battle and they will fight with honor for an honorable cause. I salute them as we all salute America today.

Very sincerely,

Stuart F. Platt

Rear Admiral, SC, USN Retired

Bill in Congress 1815

January 9, 1815

Read the first and second time and committed to a committee of the whole house on Wednesday next.

A BILL

Directing the manner of contracts and purchases in the navy department, and for promoting economy therein.

Be it enacted by the senate and house of representatives of the United States of America in congress assembled, that from after the passage of this act, all contracts for building and otherwise procuring vessels and their equipments, and supplies of every kind for the service of the navy and marine corps, where the contract or articles of equipment or supplies shall exceed the amount of one thousand dollars, the same shall be procured by public advertisements extensively circulated, requiring duplicate sealed proposals, one of which shall be immediately forwarded to the office of the accountant of the navy, for his government in the final settlement of the accounts; and no account shall be allowed for higher rates or terms, than the lowest offer thus received, unless it shall be satisfactorily shown that such could not be safely executed on the conditions hereinafter prescribed.

Sec.2 And be it further enacted, That in all proposals for contracts directed by this act, the same shall specify the quantity of the articles wanted, and the time and place of delivery; and in all cases the lowest terms shall be accepted, where sufficient security can be given for the faithful and punctual performance and no contract for services or supplies of any kind, exceeding the amount of one thousand dollars, shall be otherwise procured, excepting only when required for immediate use, and may not be delayed without injury to the service; ...

Sec.3 And be it further enacted, that from and after the _____ day of _____ next, no navy agent who has been heretofore appointed, or who may hereafter be appointed, shall act as such, until he shall have taken and subscribed the following oath, viz: I, _____ _____ do solemnly, sincerely, and truly swear, (or affirm, as the case may be,) that I will not be interested directly or indirectly in any contract I may make, or in any supplies I may purchase for, or on account of, the United States; and that in making all contracts, and in procuring all supplies, I will use my best endeavors to obtain them on the lowest terms that can be had. So help me God...

THE ARMAMENT TIDE

8

Defense Authorization Bill creating role of Competition Advocate Generals 1984

8th Congress

2nd Session

June 8, 1984, 11:05 a.m.

Page S-6873 Temp. Record

Vote No. 117

DEFENSE AUTHORIZATION/Competition Advocate

SUBJECT: Omnibus Defense Authorization Act, 1985 (S. 2723). Modified Byrd et al. amendment No. 3164, as further modified.

AMENDMENT AGREED TO, 80 - 4

SYNOPSIS: For pertinent votes on this matter, refer to the index. As reported by the Armed Services Committee, the bill includes a section establishing competition advocates to promote competition in the procurement of property and services by the armed services agencies. The Byrd et al. amendment No. 3164 would insert language, in lieu of the above section, to establish an Office of the Competition Advocate General for each branch of the armed services, to be headed by a General or an Admiral, for the purpose of reviewing every sole-source procurement over $100,000. As modified during floor debate, the Byrd et al. amendment would also provide for a Competition Advocate General for the National Aeronautics and Space Administration (NASA). In addition, the amendment as modified would provide for terms of 2 years for all Competition Advocate Generals and would allow competition requirements to be waived for national security or emergency reasons.

DEBATE: Those favoring the amendment contended:

Timely, equitable, and necessary for increased efficiency in Department of Defense procurement, the pending Byrd et al. amendment should be agreed to by the Senate. The major goal of this amendment is to put teeth in the bill's provision for competition advocates in the services and to establish competitive procurement practices in Department of Defense (DoD) purchasing procedures. The price of spare parts needed by the DoD should be determined in the crucible of competition - to the benefit of defense programs, the Federal budget, and small business. This goal will be reached in the following ways: First, an Office of the Competition Advocate General will be established. The bill provided only for the assignment of such an individual with no extraordinary power for each service. Under the amendment, the competition advocate, who will serve 2 years, will be an individual with enough clout to take on the military procurement bureaucracy. Only a high-level advocate in a protected, objective position can seriously challenge the procurement bureaucracy. Second, proposed sole-source procurements over $100,000 will be scrutinized by a competition advocate to see if others can be given a chance to bid. Further, written justification of actions inconsistent with a recommendation to compete will be required. This provision insures that a regular, independent review becomes an integral part of the procurement process. Such reviews may be waived, however, for security reasons or due to extreme emergency. Finally, each agency head will be required to periodically transmit to Congress reports of Government buys inconsistent with the competition advocate's recommendations.

Collectively, these three requirements will thrust the Federal Government into the competitive marketplace. Logic and experience teach us that the best deals are found through competitive shopping. Indeed, the DoD estimates that the Government saves, on the average, 25 percent when it buys competitively. Other studies suggest a far greater savings. The potential effects of this amendment are tremendous: In 1983, DoD spent in excess of $13 billion on spare and replenishment parts, and most of these purchases were made without the benefit of competitive bidding. It is important to note that enhanced competition would aid our Nation's small businesses and entrepreneurs. Statistically, most of the low bidders for common replacement parts and components are small businesses.

The problems caused by lack of competition in procurement procedures are endemic to the whole Government. In the corporate world, individuals are trained for 15 or 20 years to handle billions of dollars, yet we expect to teach a young officer in the Pentagon to do the same thing in only 2 or 3 years. While the provisions of this amendment will certainly help the situation, we must at the same time begin emphasizing early in military training that management of procurement is critical. In conclusion, the services must be made to understand that Congress is serious about achieving a full level of competition. A full-time Executive advocate for enforcing competition is essential if the services are to do more than merely discuss competition. Recent spare-parts procurement horror stories make it clear that we must take positive corrective action now. By accepting this amendment, we will shake up the system, jolt the ingrained habits of procurement officials, and enforce fresh and constant reviews of the adequacy of competition within the procurement process.

No arguments were voiced in opposition to the amendment.

DOCUMENTS

9

NAVY
Procurement Competition

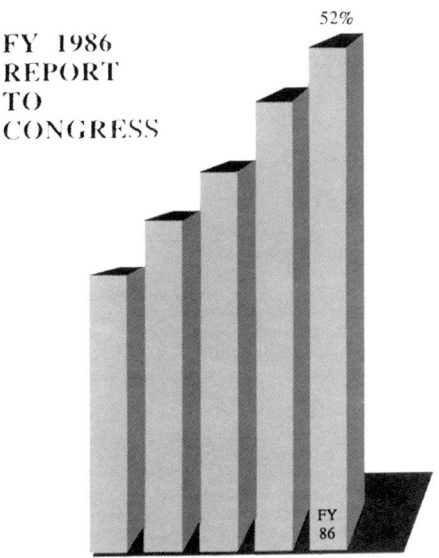

FY 1986
REPORT
TO
CONGRESS

SUCCESS through
COMMITMENT to EXCELLENCE

Office of the
Competition Advocate General
of the Navy
December 1986   Washington, D. C.

THE ARMAMENT TIDE

DEPARTMENT OF THE NAVY
THE COMPETITION ADVOCATE GENERAL
WASHINGTON, D.C. 20360-5100

12 December 1986

MEMORANDUM FOR THE SECRETARY OF DEFENSE

Via: (1) Assistant Secretary of the Navy (Shipbuilding and Logistics)
 (2) Secretary of the Navy

Subj: FY 1986 ANNUAL REPORT ON NAVY PROCUREMENT COMPETITION

Ref: (a) Competition in Contracting Act of 1984 (PL 98-369)
 (b) Defense Authorization Act of 1985 (PL 98-525)
 (c) DOD Directive 4245.9
 (d) USD&E memo of 24 Jan 1985

 I submit herewith, the Fiscal Year 1986 Annual Report on Navy Procurement Competition for inclusion in your annual report to the Congress, in compliance with the referenced statutes and Department of Defense guidance.

 A significant milestone for Navy procurement was achieved in FY 1986. For the first time since World War II, over half of Navy procurement dollars were awarded on a competitive basis. Taxpayers can be justifiably pleased with our success in controlling cost through the application of proven business practices. With the Secretary's direct guidance and the support of our civilian and military leadership, competition is now a way of life within the Navy and a fundamental guiding principle for Navy and Marine Corps acquisitions. Teamwork, among and within the Services, has also helped make our competition in contracting initiatives successful.

 The Navy set a FY 1986 competition goal of 51 percent, a 14 percent increase over its FY 1985 actual experience; the Navy exceeded that goal, competitively awarding 51.9 percent of our FY 1986 contract dollars. The Navy awarded almost 73 percent of its contract actions competitively in FY 1986, a 144 percent increase over FY 1982. This year we awarded $23.2 billion through competitive means, a major improvement over the $10 billion competitively awarded in FY 1982. Competition has been instrumental in the impressive savings achieved in naval shipbuilding, aircraft, combat systems, spare parts, and maintenance programs. Though more remains to be done, the gains already achieved bode well for the Navy and the defense establishment.

 I view that the defense industry, with a new spirit of free enterprise, has helped us set in place procurement programs that will be of lasting benefit to the nation. The economic rights that evolve and opportunities created through the increased use of competitive contracts only help to secure our free society.

Very respectfully,

STUART PLATT
Rear Admiral, SC, USN

Attachment

DOCUMENTS

FY 1986 REPORT TO CONGRESS

Table of Contents

Section	Page
Executive Summary	i
Part I. Competition Results	I-1
A. Summary Data	I-2
B. Areas of Special Interest	I-4
C. Navy Achievement Is Broadly Based	I-6
Part II. Accomplishments by Product Lines and Commodities	II-1
A. Industry Segments	II-2
B. Major Product Lines	II-10
C. Areas of Special Interest	II-11
Part III. Cornerstone of Navy Acquisition Policy	III-1
A. Introduction	III-2
B. Navy Strategy	III-2
C. Navy Acquisition Policy	III-4
D. Role of the Navy Competition Advocate General	III-6
E. Extending Competition Activities	III-8
Part IV. Overcoming Barriers to Competition	IV-1
A. The Navy Competition Advocate General Priority Objectives	IV-2
B. Other Competition-Related Initiatives	IV-14
Part V. Increased Automation Is in Our Future	V-1
A. Economic and Technical Considerations	V-2
B. Productivity Enhancement	V-3
C. Government/Industry Interrelationship	V-4
D. U.S. Worldwide Competitiveness	V-5

THE ARMAMENT TIDE

Navy Procurement Competition 1986

EXECUTIVE SUMMARY

Fiscal Year 1986 will be viewed as a **lasting benchmark** for Navy procurement. For the first time since World War II, over half our Navy procurement dollars were awarded on a competitive basis. Significant savings have been achieved in shipbuilding, aircraft, missiles, combat systems, spare parts, and maintenance programs. With Secretary Lehman's guidance and the support of our civilian and military leadership, we remain in the vanguard of obtaining the best value for the taxpayers' dollars. The Navy and Marine Corps leadership can be justifiably proud of the success of their acquisition workforce in applying sound business judgment and effective management control to the Navy contracting process. The Navy programs to improve our business management are maturing. The following charts and sections highlight the many actions accomplished throughout the Department.

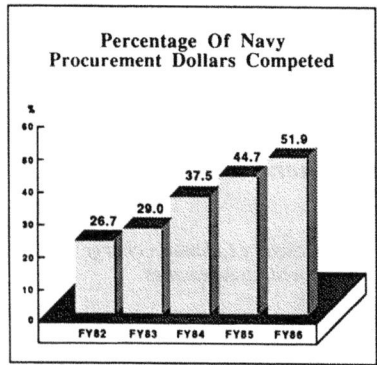

- The Navy competed more than half of its procurement dollars for the first time since World War II.

- The Navy also competed about three-fourths of its contract actions, almost two and one-half times the percentage for FY 1982.

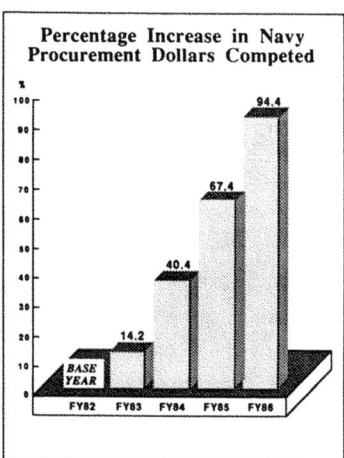

- The Navy and Marine Corps team has nearly doubled contract competition . . . a 27 percentage point increase per year since FY 1983.

DOCUMENTS

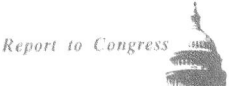 Report to Congress

A. NAVY ATTAINS ITS GOAL

The Assistant Secretary of the Navy (Shipbuilding and Logistics) set a FY 1986 competition goal of 51 percent -- a 14 percent increase over its FY 1985 actual experience. The Navy exceeded that goal, competitively awarding 51.9 percent of its FY 1986 contract dollars and almost three quarters of all procurement actions. Results were even higher for Contracted Advisory and Assistance Services (66.7 percent), higher yet for the 500 offices in the Navy field contracting system (76.6 percent), and **highest for the "grass roots" area** of small purchases (76.7 percent). Significant advances were made in spare parts procurements, with both Navy Inventory Control Points competing over 40 percent of spare parts procurements. As a result, shipboard spare parts have been procured at average prices more than 10 percent lower than the previous year for the third consecutive year.

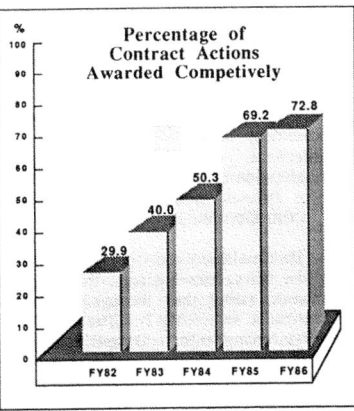

Percentage of Contract Actions Awarded Competively

B. ACROSS-THE-BOARD ACCOMPLISHMENTS FOR PRODUCT LINES

Increasing competition in Navy procurement has been accomplished across the board. In FY 1986, 108 contract actions, each over $50 million, were awarded to 39 of the "Top 100" companies; 47.8 percent of the dollar value of these contracts was awarded competitively, almost doubling the percentage of dollars awarded competitively for this category in FY 1983.

In shipbuilding, 94.7 percent of our ships under construction (83.1 percent of the dollars) was awarded competitively. Competition savings from shipbuilding amounted to $958 million in FY 1986, including $282 million in the AEGIS Cruiser Program, $225 million in the LOS ANGELES Class Attack Submarine Program, and $208 million in the Landing Helicopter Dock Ship Program. The stage for TRIDENT Program competition has been firmly set. In the ship overhaul area, all surface ship overhauls were solicited and competitively awarded. We were able to compete all of the private sector ballistic missile submarine overhauls that previously would have been awarded on a sole source basis. Seventeen ships were competed under the test of public and private shipyard competitions begun in FY 1985.

Competitive awards continue to increase in the aerospace industry. Nearly one-third of all procurement dollars awarded by the Naval Air Systems Command were

iii

THE ARMAMENT TIDE

Navy Procurement Competition 1986

competed in FY 1986 -- two and one-half times more than the FY 1983 figure. A competitive award was made for the V-22 (Osprey) aircraft full scale engineering development (FSED), and the short-range Remotely Piloted Vehicle was awarded competitively.

Second production sources were selected for several top-line weapons systems, including the Phoenix missile, thus ending 15 years of sole source procurements. Additionally, the Navy ADP Selection Office competed over 99 percent of its procurement dollars for non-tactical automated data processing hardware, software, maintenance, and services, setting a new standard for the Services' central ADP-selection offices.

C. CORNERSTONE OF NAVY ACQUISITION POLICY

The last three and one half years have been marked by a profound change in the way the Navy does business; this change has been brought about more by people and teamwork, rather than through regulation (as was the case with the Armed Services Procurement Act of 1947). The Competition in Contracting Act has played a meaningful and reinforcing role. Competition is now a well-established foundation for acquisition of all systems and equipment. Our Navy's transformation to competition has brought with it a corresponding structural change in the defense industry. Many corporations have been alerted to the new business opportunities competition makes available. A new spirit of free enterprise has been kindled; we strive to intelligently apply our policy of affordability through competition, while continuing to emphasize accountability, discipline, and quality. These basic concepts are woven throughout the entire Navy chain of command and, by linking policy with execution, we have sought both new and bold solutions to better manage the acquisition process.

Navy policy requires dual sources for the development and production of weapons systems provided it is economically feasible. During FSED, two contractors will develop a system to the same design. These two sources will then compete against each other during the production phase. Overall, the Navy strives to acquire equipment having the requisite quality and value for our active duty and reserve forces. During FY 1986, detailed memoranda and communiques emphasizing the **best value concept** were issued, providing guidance to ensure quality and addressing contractor performance in evaluating Navy awards. The Navy is also using "should cost" and other techniques to help our buyers deal with sole source situations where the marketplace forces of competition are not present.

The Navy acquisition organization is highlighted by clear lines of authority. Responsibility and accountability are vested in system command leadership, program managers, and selected field activity commanders with oversight control exercised through technical and management milestone reviews by the Navy's senior executives. Review and oversight of competition business planning is carried out, in part, by the Office of the Navy Competition Advocate General and a network of over 250 Navy and Marine Corps Competition Advocates at hardware systems commands and field activities.

iv

DOCUMENTS

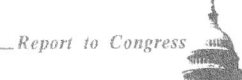 _Report to Congress_

The Office of the Navy Competition Advocate General, as part of the Navy's acquisition decision team, reviewed over 325 top-level acquisition plans during FY 1986.

The Navy competition infrastructure is in place and working well. Competition in the marketplace provides a stronger industrial base and protects the economic opportunities and rights that secure our free society.

D. OVERCOMING BARRIERS TO COMPETITION

During FY 1986, the Navy Competition Advocate General established a Priority Objectives Program to address barriers to competition; this program consists of 34 initiatives.

Under the Navy Competition Priority Objectives Program, a wide variety of analyses and actions are in progress to eliminate or reduce barriers that hamper the intelligent application of competition. Competition planning at the systems command level was emphasized to develop realistic command competition goals and convert sole source procurements to competition where possible. Particular attention was again given Acquisition Plans and Justification and Approvals for sole source procurements greater than $10 million; this important effort will continue in FY 1987. Many other planning activities are underway. For example, the Naval Sea Systems Command is developing a five-year market research plan projecting its Contracted Advisory and Assistance Services requirements in order to better inform potential bidders. NAVSEA has also set up a 24-hour telephone answering service to provide information on solicitations.

Studies of factors governing the decision to second source production of weapons systems were conducted in FY 1986. Work continues on developing an improved computer-based cost and benefit model to assist in realistically conducting economic analyses.

The Office of the Navy Competition Advocate General continued its communications role in FY 1986 by hosting a competition symposium, increasing dialogue with industry, and publishing guidance to Navy and Marine Corps Competition Advocates. Studies involving various elements of procurement workload, the contracting process, and administrative leadtime are underway and will continue into FY 1987. Specialized training programs were developed and conducted in production competition, market research, breakout analysis, and pricing reviews; these programs will also continue.

Excellent progress has been made in acquisition streamlining, spare parts breakout programs, and special technology productivity programs. To facilitate industry participation in Navy acquisition planning, we have increased the use of draft requests for proposals, promoted the use of performance specifications, enlisted

THE ARMAMENT TIDE

Navy Procurement Competition 1986

industry participation in setting initial requirements, and conducted awareness briefings with associations and industrial companies. Under project BOSS (Buy Our Spares Smart), a "Spare Parts Annex" to all Navy Acquisition Plans addresses actions taken to seek, promote, and sustain early competition in spare parts acquisition. A special program to review technical documentation for all major programs is underway, and a breakout deficiency tracking system has been initiated.

The Navy has started a number of technology development programs which will improve industry's ability to compete for spare parts contracts. Under the coordination of the Naval Supply Systems Command, we are developing a Navy Standard Technical Information System which will provide technical data to bidders in a more timely fashion; the Rapid Acquisition of Manufactured Parts program will reduce the cost of low-volume spares production; and an Automated Procurement and Accounting Data Entry system will provide contracting officers with more timely and complete procurement data.

The Navy's wide range of commitments to improve its business planning and competitive practices is being carried out smartly.

E. **LOOKING TO THE FUTURE**

We have now begun to rapidly capitalize on our business philosophy through the management infrastructure the Navy has established. To keep pace with these improvements, the Navy acquisition team will look to new methods to replace and streamline the paper-intensive procurement process. The advancement of Navy procurement automation, if left unattended, will be a major barrier to successful competition in the future. We will work to set in place expert systems to reduce lead times in the procurement process. Also, steps that capitalize upon defense industry productivity and Navy competition will further improve the acquisition process. We in the armed services, through interactions with a segment of the U.S. industrial base, increasingly find ourselves in a position to aid our nation's worldwide competitiveness.

The Navy Procurement Competition Program, which started over three years ago, is working well. Competition in contracting has set the stage for an affordable 600 ship Naval Force. This program has clearly benefitted the taxpayers and our Navy.

STUART PLATT
Rear Admiral, SC, USN
Competition Advocate General

vi

10

Executive Order for creation of Director of Naval Nuclear Propulsion Program

DATE: 02-01-82

32 – National Defense Subchapter A – General Naval Nuclear Propulsion Program

By the authority vested in me as President and as Commander in Chief of the Armed Forces of the United States of America, with recognition of the crucial importance to national security of the Naval Nuclear Propulsion Program, and for the purpose of preserving the basic structure, policies, and practices developed for this Program in the past and assuring that the Program will continue to function with excellence, it is hereby ordered as follows:

Section 1. The Naval Nuclear Propulsion Program is an integrated program carried out by two organizational units, one in the Department of Energy and the other in the Department of the Navy.

Sec. 2. Both organizational units shall be headed by the same individual so that the activities of each may continue in practice under common management. This individual shall direct the Naval Nuclear Propulsion Program in both departments. The director shall be qualified by reason of technical background and experience in naval nuclear propulsion. The director may be either a civilian or an officer of the United States Navy, active or retired.

Sec. 3. The Secretary of the Navy (through the Secretary of Defense) and the Secretary of Energy shall obtain the approval of the President to appoint the director of the Naval Nuclear Propulsion Program for their respective Departments. The director shall be appointed to serve a term of eight years, except that the Secretary of Energy and the Secretary of the Navy may, with mutual concurrence, terminate or extend the term of the respective appointments.

Sec. 4. An officer of the United States Navy appointed as director shall be nominated for the grade of Admiral. A civilian serving as director shall be compensated at a rate to be specified at the time of appointment.

Sec. 5. Within the Department of Energy, the Secretary of Energy shall assign to the director the responsibility of performing the functions of the Division of Naval Reactors transferred to the Department of Energy by Section 309(a) of the Department of Energy Organization Act (42 U.S.C. 7158), including assigned civilian power reactor programs, and any naval nuclear propulsion functions of the Department of Energy, including:

(a) direct supervision over the Bettis and Knolls Atomic Power Laboratories, the Expended Core Facility and naval reactor prototype plants;

(b) research, development, design, acquisition, specification, construction, inspection, installation, certification, testing, overhaul, refueling, operating practices and procedures, maintenance, supply support, and ultimate disposition, of naval nuclear propulsion plants, including components thereof, and any special maintenance and service facilities related thereto;

(c) the safety of reactors and associated naval nuclear propulsion plants, and control of radiation and radioactivity associated with naval nuclear propulsion activities, including prescribing and enforcing standards and regulations for these areas as they affect the environment and the safety and health of workers, operators, and the general public;

(d) training, including training conducted at the naval prototype reactors of the Department of Energy, and assistance and concurrence in the selection, training, qualification, and assignment of personnel reporting to the director and of personnel who supervise, operate, or maintain naval nuclear propulsion plants; and

(e) administration of the Naval Nuclear Propulsion Program, including oversight of program support in areas such as security, nuclear safeguards and transportation, public information, procurement, logistics, and fiscal management.

Sec. 6. Within the Department of Energy, the director shall report to the Secretary of Energy, through the Assistant Secretary assigned nuclear energy functions and shall serve as a Deputy Assistant Secretary. The director shall have direct access to the Secretary of Energy and other senior officials in the Department of Energy concerning naval nuclear propulsion matters, and to all other personnel who supervise, operate or maintain naval nuclear propulsion plants and support facilities for the Department of Energy.

Sec. 7. Within the Department of the Navy, the Secretary of the Navy shall assign to the director responsibility to supervise all technical aspects of the Navy's nuclear propulsion work, including:

(a) research, development, design, procurement, specification, construction, inspection, installation, certification, testing, overhaul, refueling, operating practices and procedures, maintenance, supply support, and ultimate disposition, of naval nuclear propulsion plants, including components thereof, and any special maintenance and service facilities related thereto; and

(b) training programs, including Nuclear Power Schools of the Navy, and assistance and concurrence in the selection, training, qualification, and assignment of personnel reporting to the director and of Government personnel who supervise, operate, or maintain naval nuclear propulsion plants.

Sec. 8. Within the Department of the Navy, the Secretary of the Navy shall assign to the director responsibility within the Navy for:

(a) the safety of reactors and associated naval nuclear propulsion plants, and control of radiation and radioactivity associated with naval nuclear propulsion activities, including prescribing and enforcing standards and regulations for these areas as they affect the environment and the safety and health of workers, operators, and the general public.

(b) administration of the Naval Nuclear Propulsion Program, including oversight of program support in areas such as security, nuclear safeguards and transportation, public information, procurement, logistics, and fiscal management.

Sec. 9. In addition to any other organizational assignments within the Department of the Navy, the director shall report directly to the Chief of Naval Operations. The director shall have direct access to the Secretary of the Navy and other senior officials in the Department of the Navy concerning naval nuclear propulsion matters, and to all other Government personnel who supervise, operate, or maintain naval nuclear propulsion plants and support facilities.

Sec. 10. This Order is effective on February 1, 1982.

The provisions of Executive Order 12344 of Feb. 1, 1982, appear at 47 FR 4979, 3 CFR, 1982 Comp., p. 128, unless otherwise noted.

THE ARMAMENT TIDE

11

Inscription on the Statue of Liberty at Ellis Island

The Statue of Liberty was given as a gift to America from the people of France. The poet, Emma Lazarus, named the Statue "Mother of Exiles." Her poem "The New Colossus" from which the line "...give me your tired, your poor, your huddled masses yearning to breathe free" is taken, is inscribed on a plaque at the foot of the statue.

THE NEW COLOSSUS - EMMA LAZARUS - 1883

Not like the brazen giant of Greek fame

With conquering limbs astride from land to land;

Here at our sea-washed, sunset gates shall stand

A mighty woman with a torch, whose flame

Is the imprisoned lightning, and her name

Mother of Exiles. From her beacon-hand

Glows world-wide welcome; her mild eyes command

The air-bridged harbor that twin cities frame,

"Keep, ancient lands, your storied pomp!" cries she

With silent lips. "Give me your tired, your poor,

Your huddled masses yearning to breathe free,

The wretched refuse of your teeming shore,

Send these, the homeless, tempest-tossed to me,

I lift my lamp beside the golden door!"

APPENDIX A
World-wide GDP
CVN Fact Sheet

THE ARMAMENT TIDE

WORLD COMPARISON OF DEFENSE SPENDING AS A PERCENTAGE OF GDP

Azerbaijan: 2.6% $121m (FY99)
Afghanistan: NA% $NA
Albania: 1.5% $42m (FY99)
Algeria: 4.1% $1.87b (FY99)
Angola: 22% $1.2b (1999
Antigua: NA% $NA
Argentina: 1.3% $4.3b (FY99)
Armenia: 4% $75m (FY99)
Australia: 1.9% $6.9b (FY98/99)
Austria: 1.2% $1.7b (FY98)
Azerbaijan: 2.6% $121m (FY99)
Bahamas: NA% $NA
Bahrain: 5.2% $318m (FY99)
Bangladesh: 1.8% $559m (FY96/97)
Barbados: NA% $NA
Belarus: 1.2% $156m (FY98)
Belgium: 1.2% $2.5b (FY99)
Belize: 2.4% $17m (FY98/99)
Benin: 1.2% $27m (FY96)
Bermuda: NA% $NA
Bhutan: NA% $NA
Bolivia: 1.8% $147m (FY99)
Bosnia Herzegovina: NA% $NA
Botswana: 1.2% $61m (FY99)
Brazil: 1.9% $13.4b (FY99)
Brunei: 5.1% $343m (FY98)
Bulgaria: 2.4% $344m (FY00)
Burkina Faso: 2% $66m (FY96)

Burma: 2.1% $39m (FY97/98
Burundi: 6.1% $57m (FY97)
Cambodia: 3% $112m (FY01 est.)
Cameroon: 1.4% $118.6m (FY98/99)
Canada: 1.3% $7.5b (FY00/01)
Cape Verde: 1.8% $4m (FY96)
Central Afr. Rep.: 2.2% $29m (FY96)
Chad: 3.5% $39m (FY96)
Chile: 3.1% $2.5b (FY99)
China: 1.2% $12.6b* (FY99)
Columbia: 3.4% $3b (FY00)
Comoros: NA% $NA
Dem. Rep. Congo: 4.6% $250m (FY97)
Rep. Congo: 3.8% $110m (FY93)
Costa Rica: 1.6% $69m (FY99)
Cote d'Ivoire: 1% $94m (FY96)
Croatia: 3.8% $575m (2000)
Cuba: roughly 4% $NA (FY95 est.)
Cyprus: 4.2% $370m (FY00)
Czech Republic: 2.2% $1.2b (FY01)
Denmark: 1.4% $2.47b (FY99)
Djibouti: 4.5% $23m (FY97)
Dominica: NA% $NA

APPENDIX A

Dominican Rep.: 1.1% $180m (FY98)
Ecuador: 3.4% $720m (FY98)
Egypt: 4.1% $4.04b (FY99/00)
El Salvador: 0.7% $112m (FY99)
Equatorial Guinea: 0.6% $3m (FY97/98)
Eritrea: 29.4% $160m (2000 est.)
Estonia: 1.2% $70m (FY99)
Ethiopia: 2.5% $138m (FY98/99)
Falkland Islands: NA% $NA
Faroe Islands: NA% $NA
Fiji: 1.1% $24m (FY98)
Finland: 2% $1.8b (FY98)
France: 2.5% $39.83b (FY97)
French Guiana: NA% $NA
Gabon: 1.6% $91m (FY96)
Gambia: 2% $2.6m (FY96/97)
Gaza Strip: NA% $NA
Georgia: 0.59% $23m (FY00)
Germany: 1.5% $32.8b (FY00)
Ghana: 0.7% $53m (FY99)
Greece: 4.91% $6.12b (FY99/00 est.)
Grenada: NA% $NA
Guatemala: 0.6% $120m (FY99)
Guinea: 1.4% $56m (FY96)
Guinea-Bissau: 2.8% $8m (FY96)
Guyana: 1.7% $7m (FY94)
Haiti: NA% $NA
Honduras: 0.6% $35m (FY99
Hong Kong: NA% $NA
Hungary: 1.6% $822m (FY00)
India: 2.5% $13.02b (FY00)
Indonesia: 1.3% $1b (FY98/99)
Iran: 2.9% $5.787b (FY98/99)
Iraq: NA% $NA
Ireland: 0.75% $738m (2001 est.)
Israel: 9.4% $8.7b (FY99)
Italy: 1.7% $20.7b (FY00/01)
Jamaica: NA% $30m (FY95/96 est.)
Japan: 0.96% $43b (FY01)
Jordan: 7.8% $608.9m (FY98/99)
Kazakhstan: 1.5% $322m (FY99)
Kenya: 1.9% $197m (FY98/99)
Kiribati: NA% $NA
Korea, North: 25%-33% $3.7b - $4.9b (FY98 est.)
Korea, South: 3.2% $12b (FY98/99)
Kuwait: 8.7% $1.9b (FY00/01)
Kyrgyzstan: 1% $12m (FY99)
Laos: 4.2% (FY96/97) $55m (FY98)
Latvia: 0.9% $60m (FY99)
Lebanon: 4.8% $343m (FY99/00)
Lesotho: NA% $34m (FY99)

THE ARMAMENT TIDE

Liberia: 2% $1m (FY98)
Libya: 3.9% $1.3b (FY99/00)
Lithuania: 1.66% $181m (FY00)
Luxembourg: 1% $131m (FY98/99)
Macedonia: 2.17% $76.3m (FY00/01)
Madagascar: 1% $29m (FY94)
Malawi: 0.76% $9.5m (FY00/01)
Malaysia: 2.03% $1.69b (FY00)
Maldives: NA% $NA
Mali: 2% $49m (FY96)
Malta: 1.7% $60m (2000)
Marshall Islands: NA% $NA
Mauritania: 2.7% $41m (FY97/98)
Mauritius: 0.3% $11m (FY97/98)
Mexico: 1% $4b (FY99)
Moldova: 1% $6m (FY99)
Mongolia: 2.3% $25.5m (FY01)
Morocco: 4% $1.4b (FY99/00)
Mozambique: 1% $35.1m (2000 est.)
Namibia: 2.6% (FY97/98) $104.4m (FY01)
Iraq: NA% $NA
Namibia: 2.6% (FY97/98) $104.4m (FY01)
Nauru: NA% $NA
Nepal: 0.9% $44m (FY96/97)

Netherlands: 1.5% $6.5b (FY00/01 est.)
New Caledonia: 5.3% $192.3m (1996)
New Zealand: 1.1% $883m (FY 97/98)
Nicaragua: 1.2% $26m (FY98)
Niger: 1.1% $20m (FY96)
Nigeria: 10% $360m (FY00)
Norway: 2.1% $3.113b (FY98)
Oman: 13% $2.4b (FY00)
Pakistan: 3.9% $2.435b (FY99/00)
Palau: NA% $NA
Panama: 1.3% $128m (FY99)
Papua New Guinea: 1% $42m (FY98)
Paraguay: 1.4% $125m (FY98)
Peru: 1.9% $1b (FY00)
Philippines: 1.5% $995m (FY98)
Poland: 1.95% $3.17b (FY00)
Portugal: 2.6% $2.458b (FY97)
Qatar: 10% $723m (FY00/01)
Romania: 2.2% $720m (FY00)
Russia: NA% $NA
Rwanda: 3.2% $58m (FY01)
Saint Kitts and Nevis: NA% $NA
Saint Lucia: 2% $5m (FY91/92)
St. Vincent and Grenadines: NA% $NA
Samoa: NA% $NA
San Marino: NA% $NA

APPENDIX A

Sao Tome and Principe: 1.5% $1m (FY94)
Saudi Arabia: 13% $18.3b (FY00)
Senegal: 1.4% $68m (FY97)
Seychelles: 2.8% $13m (FY93)
Sierra Leone: 2% $46m (FY96/97)
Singapore: 4.5% $5b (FY00/01 est.)
Slovakia: 1.71% $380m (FY00)
Slovenia: 1.7% $370m (FY00)
Solomon Islands: NA% $NA
Somalia: NA% $NA
South Africa: 1.5% (FY99/00) $2b (FY00/01)
Spain: 1.1% $6b (FY97)
Sri Lanka: 4.2% $719m (FY98)
Sudan: NA% $550m (FY98)
Suriname: 1.6% $8.5m (FY97 est.)
Swaziland: 4.75% $19.198m (FY00/01)
Sweden: 2.1% $5b (FY98)
Switzerland: 1.2% $3.1b (FY98)
Syria: 5.9% $921m* (FY98)
Taiwan: 2.8% $8.042b (FY98/99)
Tajikistan: 1.8% $17m (FY97)
Tanzania: 0.2% $21m (FY98/99)
Thailand: 1.4% $1.775b (FY00)
Togo: 2% $27m (FY96)
Tonga: NA% $NA
Trinidad and Tobago: NA% $83m (FY94)
Tunisia: 1.5% $356m (FY99)
Turkey: 5.6% $10.6b (FY99)
Turkmenistan: 3.4% $90m (FY99)
Tuvalu: NA% $NA
Uganda: 1.9% $95m (FY98/99)
Ukraine: 1.4% $500m (FY99)
United Arab Emirates: 3.1% $1.6b (FY00)
United Kingdom: 2.7% $36.884b (FY97)
U.S.A.: 3.2% $276.7b (FY99 est.)
Uruguay: 0.9% $172m (FY98)
Uzbekistan: 2% $200m (FY97)
Vanuatu: NA% $NA
Venezuela: 0.9% $934m (FY99)
Vietnam 2.5% $650 (FY98)
West Bank: NA% $NA

THE ARMAMENT TIDE

NUCLEAR POWERED AIRCRAFT CARRIER CVN FACT SHEET

BUILDER
Newport News Shipbuilding and Dry Dock Company, Newport News, Virginia

COST (1980)
Approximately 3.9 billion dollars

PROPULSION.
Nuclear, fuel for 20 years of normal operation

SPEED
Thirty plus knots

NUMBER OF REACTORS
two

LENGTH OVER FLIGHT DECK
1,092 feet

BREADTH OF FLIGHT DECK
252 feet

AREA OF FLIGHT DECK
4.5 Acres

NUMBER OF AIRCRAFT ELEVATORS
four

HEIGHT, KEEL TO MAST
244 feet (equal to a 24 story building)

ANCHORS
two at 30 tons each

WEIGHT OF ANCHOR CHAIN LINKS
360 pounds each

COMBAT LOAD DISPLACEMENT
95,000 tons

PROPELLERS
Four, at 21 feet high and 66,200 pounds each

NUMBER OF TELEPHONES
over 2,000

APPENDIX A

WEIGHT OF RUDDERS
    65.5 tons each

DAILY CAPACITY DISTILLING PLANTS
    400,000 gallons of fresh water
    (enough for the daily needs of more than 2,000 homes)

APPENDIX B
Business Principles by Topic
Business Principles by Chapter

THE ARMAMENT TIDE

BUSINESS PRINCIPLES BY TOPIC

THE PURPOSE OF THE BOOK
DEFENSE POLICY RECOMMENDATIONS
BUSINESS AND MANAGEMENT PRINCIPLES
SPECIFIC TACTICS, COLLECTED BY TOPIC
ORGANIZATIONAL CHANGES FOR DOD

THE PURPOSE OF THE BOOK

The U.S. is now under pressure to purchase large amounts of military equipment rapidly. The defense establishment has not experienced this since the 1980s under President Reagan. The purpose of this book is to remind America how to buy smart.

DEFENSE POLICY RECOMMENDATIONS

Professional defense procurement is a keystone to rearming America effectively. Spending correctly can be more important than how much is spent.
- "It isn't just about numbers. In addition to buying enough ships and aircraft, we must buy the right ships and aircraft, with the right capabilities for our future fleet." – Adm. Vern Clark, CNO, April 2002, Senate Sea Power Subcommittee hearing.
- The Navy has indicated that investment in technology and promotion of better business practices are listed alongside readiness and personnel as building blocks of the future.

Homeland defense will focus on prevention and reaction to attack.
- Prevention priorities should be border security and intelligence gathering on the ground.
- Reaction must provide for the safety of the first responders such as local police, fire departments, and specialist elements of the National Guard and military.
- Homeland defense procurement should be treated like the

big business that it is.
- Business and community leaders must enable citizens to contribute via National Guard, Reserves, or USA Freedom Corp.

Reform of defense procurement for the military and for homeland defense is not an issue for tomorrow. Each year that we delay, we needlessly put billions of dollars, and unknown lives, at risk.

BUSINESS AND MANAGEMENT PRINCIPLES

The changed defense industry environment requires change in defense procurement.
- Competition between vendors is good for buyers, and must be encouraged. The threat of a new competitor can be almost as good as the real thing.
- The defense industry is becoming less competitive. This direction must be changed.
- Defense buyers need to become the equals of the large, sophisticated, politically-influential vendors.
- Capabilities of both individuals and teams must be improved.
- Political endurance and longevity are important factors, and must be encouraged.
- Knowledge of vendor operations and economics is important. Procurement professionals must deeply understand the business operations of the vendors to avoid becoming dependent on the vendors themselves for understanding cost and quality.
- Substantial sophistication and experience are needed either for practical input to vendors or critical interpretation of vendors.

It is important to maintain a market system.
- Get the benefits of competition without the problems of micro-management.

- Maintain market dynamics. Get the most out of market economy dynamics: don't move to quasi-nationalized structures.
- Support competition, while defending business principles such as intellectual property rights. This keeps all parties honest as well as letting vendors know where they stand at all times.
- Don't over-regulate, violate vendor property rights, violate contracts, or nationalize

Procurement practices need to be open and transparent, with bright lines identifying the authority, or absence thereof, of the procurement professionals.

- Processes should be understood by all and agreed to in advance by vendors, to reduce surprises for everyone and blunt political favoritism.
- Distinct practices for distinct phases and sizes of programs are necessary.
- Collaboration, with military as prime contractor for research and early development, must be encouraged.
- Contract competition must be managed, with major subcontracting work given to losing team for late development and volume production.
- Contracting teams must be different than the collaborative team used in R&D phases.
- Costs need to be managed down through production phases.

SPECIFIC TACTICS, COLLECTED BY TOPIC

COMPETITION MANAGEMENT

Manage procurement to increase competition.
- Avoid sole-source contracts, when possible.
- Increase the number of competitors. More effort for larger contracts.
- Increase the degree of direct competition between vendors.

- Encourage and enable existing vendors to compete in new categories they are not in but that are closely related to their expertise, especially where there is a sole source.
- Remember that the threat of new competition is almost as good as the real thing.
- Don't push losing vendors out of competitive position for the future.
- Specifically protect losing competitors where their loss would decrease future competition.
- Provide losers (especially on capital projects) with adequate sub-contracting to remain viable until the next generation of projects.
- Where there is a single asset such as software or data rights or another indivisible asset that is the necessary foundation of competition, the government should maintain ownership or control. This enables government to provide access to the asset to enable new entrants to compete.
- Measure for results. The mere reporting of those results will cause improvements.
- Buy off the shelf.

MANAGEMENT OF MAJOR PROGRAMS

Manage major programs for stability.
- Multi-decade projects and programs that evolve over multiple generations of technology need to have long-term stability.
- Programs must be managed to allow for long-term focus and accountability.
- Provisions must be made for career advancement for procurement officers.
- People must be rewarded for quality service. There is nothing wrong with promoting a person several times in the same job so long as s/he serves our country well.

- People need to stay in place for a long time, so they can learn jobs and also live with consequences of successes and mistakes.
- Natural career paths into and out of procurement must be available.
- Budgets and plans must stretch across political cycles.
- Personnel should be the best of civilian and military people.

RESEARCH AND DEVELOPMENT PROJECTS

The military should be the prime contractor during the early stages, shifting more responsibility to vendors as program matures.
- Allows management judgment and congressional oversight to react to changes, reducing the impact of change on contracts.
- Speeds overall development, reduce costs from change orders and inter-vendor conflicts.
- Matches uncertain nature of developing new technology.
- Obtain live field data from most realistic real world situations as possible, early in the research phase.
- Work collaboratively with vendors during basic research and early development.
- Use overseas ideas for new weapons. There are some very good developments going on in Europe.
- Use separate teams from contracting for R&D and collaborating with chosen vendor(s).

BIG TICKET PROJECTS

- Use a different team for procuring capital goods than was used for R&D stage of the program. The collaborative military team will be too close to some vendors for dispassionate supervision of vendors.
- Obtain multiple bids from capable vendors, with checkpoints over time where further phases of program can be

APPENDIX B

awarded to different vendors.
- Separate economics of scaling up to produce from economics of actual production. Slack production capacity often has value, whether it is used or not. But it should not be paid for if it is not wanted.
- Construct one big deck carrier per presidential term.
- Manage for careful and accurate long-term planning by Chief of Naval Operations.
- Create congressional will to see through long term projects.
- Create a strong industrial base.
- Structure terms of contracts for capital ships to avoid frivolous change orders or otherwise gain "after the fact" profits.
- Support new entrants to capital projects.

BELTS, BULLETS & BEANS PROJECTS
- Delegate vendor management downward to mid-level and junior procurement professionals. Allow mid-level and junior officers to drive ideas. Ask for ideas; review them; then turn them loose to get things done.
- Split contracts between vendors, with larger shares going to best vendors.
- Encourage low operating volatility, using higher inventory as a buffer (U.S. government has a low carrying cost of inventory).
- Use production data (collected as part of contract) to empower competition.

CONTRACT MANAGEMENT (ANY SIZE OR TYPE OF CONTRACT)
- Close contracts. Left open they are a source of potential liability.
- Identify and manage contractors' financial and contract performance. Keep ahead of them; don't let them fester.
- Keep programs to basics; avoid frills. Soldiers with only high school education and quality military training will use them

under the harshest of combat conditions.
- Develop an exit strategy for purchase cancellations. Without an exit strategy, the cost of getting out of a contract can end up as expensive as building the system – except the costly cancellation does not add to our nation's defense.
- Be specific about non-performance terms, with allowance for cancellation at the major points where such decisions are typically taken (Component Delivery, Pre-production Prototype Delivery, Initial run completion, etc.).
- Begin and end contracts at the negotiating table. Cancellation should be done by the contracting officer responsible, by the Service involved, with company officials taking a role

ORGANIZATIONAL CHANGES

Homeland Defense procurement must be professional and centralized.
- Centralize procurement so that equipment can be used across the country. Assure coordinated spending, interoperability and compatibility. (Individual naval base commanders don't buy their own ships!!)
- Control money tightly, and measure results.
- Understand the problem first, then what constitutes a response, before spending money on it.
- Have a tight, logical connection between the problem, the response, the dollars, and the contract.
- Avoid stopgap spending that provides sense of "doing something" but doesn't move forward.
- Expand ranks of highly-mobile, anti-terrorist trained soldiers. Expand their training to include working with local Emergency Response Units and in working at borders and ports.

Create a single DoD procurement authority that enjoys the sup-

port of Congress and the different branches of the military.
- Bring the separate efforts of the services into a single operating unit such as The Defense Material Procurement Agency.
- Put a newly created Chief of Defense Material Procurement at its head.
- Make the Chief of Defense Material Procurement office that of a Four Star Officer
- Give the Chief of Defense Material Procurement the same rank as Current CinCs, to ensure ability to carry forward his command.
- Make the Chief of Defense Material Procurement a six-year posting to ensure continuity through changes in Administrations.
- Focus on acquisitions with a face value of $250 million or more to limit the scope of the program to those big-ticket, long-term systems that are so vital to our nation's defense.

MAINTAIN CONGRESSIONAL OVERSIGHT OF PROCUREMENT
- Rationalize the congressional oversight process, but maintain screening process that provides national debate on major purchases.
- Back up recommendations to Congress on whether something is needed or not needed with hard answers to the dollar.

ATTRACT AND RETAIN THE BEST PEOPLE.

We can't afford to lose them at the mid-point of their careers, just as their accumulated knowledge and experience become most valuable. We cannot expect people to be motivated by purely financial considerations; they must have a deep sense of duty and patriotism. However, financial benefits should also accrue to them. Compensation / Retention Financial tactics could include:

THE ARMAMENT TIDE

- Annual pay/benefit/pension increases to reward long term service.
- Some exemption on income taxes for civilian and military personnel.

BUSINESS PRINCIPLES BY CHAPTER

CHAPTER 1

THE FLOOD TIDE OF REARMAMENT
BUSINESS PRINCIPLES

Competition is good. The more sellers competing for your business, the likelier you are to get a good deal – no matter what you are buying. Concentration among defense vendors wasn't a problem in the 1970s or even into the late 1980s. We bought at low and/or declining levels, allowing the Pentagon to extract excellent deals and manage around most problems. This is simple, even classic, economics. Supply meets demand, and whichever there is less of gets the best deal.

Consider a buyer, ranking sellers by declining value of a desired product, from best value to worst value. If there are a large number of sellers, there will be some parts of the list where the differences of value are fairly small. In this part of the "supply curve," competition is very intense, forcing sellers to do their best to move up the list. However, if there are only a few sellers, there can be large gaps in value between consecutive sellers on the list. What economists might call a jagged supply curve, this creates "natural winners." If there is a large gap between the first and second, it is often not worth the time and money for the number two vendor to try to move up. When this pattern repeats itself for different product types, with different winners, the result is a whole series of single-sources for critical goods. This is a real risk for the current rearmament task, with the defense vendor market now much more concentrated than it was the last time (1980s) we expanded military purchasing.

The correct reaction is to change how the demand side behaves. Unlike most industries, defense purchasing is influenced strongly by congress, DoD, lobbying, national debates, and the many purchasing points in the Pentagon, along with the will of the public to be taxed – the final criterion. Moreover, today price has ceased to be the top issue. We are in a war and we want to win it. Price is important, but results matter more (a fact likely not missed by the vendors). Many of the factors that complicate defense purchasing are necessary or even desirable. Fragmentation and attenuation of purchasing power is decidedly not. When a jagged supply curve meets a smooth demand curve, some buyers get good value, and others do not. They are forced to choose from the second or third supplier, which often have dramatically lower overall value. Moreover, with smaller individual budgets, a large number of buyers have great trouble convincing a lower tier seller that an investment to improve could be rewarded by a large contract. However, consolidating purchasing makes the demand curve look rather more like the supply curve. With greatly reduced Pentagon purchasing entities, the Pentagon would have negotiating power commensurate with the top vendors and still enough carrots available to motivate the less capable vendors to improve their capabilities.

To achieve good value for taxpayers' money, procurement should be managed to increase competition. We learned this lesson in the 1980s from the many capital ships and systems that were bought under multi-year contracts, sourced from a single supplier. From a strictly business point of view most CEOs would be alarmed by the implications to the shareholders of sole source supply to the company's production line. We were as well, and used basic procurement (demand) management principles to fix the problems.

CHAPTER 2

BIG SHIPS, BIG COSTS, BIG SAVINGS

BUSINESS PRINCIPLES

Getting the most out of vendors requires that procurement professionals have a deep understanding of the business of their vendors. In the instance of CVN procurement, substantial savings were achieved by efficiently utilizing the specialized (but very expensive to start up and shut down) dry-docks needed for construction. A strong understanding of what drives vendor costs and profits is needed to avoid becoming dependent on the vendors themselves, in order to understand what makes sense in terms of cost and quality. This sophistication is needed whether to contribute practical input or to watch for inappropriate vendor practices.

This depth of familiarity with CVN construction allowed us to create a credible threat of producing at Bremerton – creating price pressure on the incumbent. Without this, Newport News wouldn't have pressured itself as hard to develop technology and to compete better on cost.

CHAPTER 3

THE GENERAL ELECTRIC F404 –
THE ENGINE THAT KEEPS GOING, AND GOING...

BUSINESS PRINCIPLES

Big Business is Big Business. Rewards are immense to winners who take all. Competition is good; but, pushed too far, it can decrease competition in the future. You have to leave the market available to work for you again and again, or you lose the future ongoing benefits of cost and technological competition.

APPENDIX B

Eventually, this leads to paying too much for technology that has become stagnant. Smart purchasing includes protecting losing competitors where their loss would decrease future competition. In many cases these companies are a significant national asset.

With jet engines, manufacturing is extremely sophisticated. Parts must be manufactured to very high tolerances and be of very high quality construction; they are expected to experience great extremes temperature and pressure for extended periods without a single failure. The skills needed to be able to produce fighter jet engines are expensive and rapidly dissipate without a production contract. Also, each type of engine is different, with advancing technologies often requiring new expertise to be integrated into the existing skill base in order for a manufacturer to even bid on a contract. In business terms, the jet engine business has much higher entry barriers than exit barriers. This creates a natural tendency for vendors to leave the industry when faced with adversity or declining orders, a tendency that smart procurement will help to balance.

However, simply wanting a second (or a third) source is only the beginning. This desire does not change the facts of the entry barriers; it only allows for the sharing of the costs and management of the entry barriers. With the F404, we recognized that one key factor was technical knowledge. So, we had a practical problem of how to effectively transfer that technical knowledge, owned by the U.S. taxpayer, to the prospective vendor, while protecting the technical knowledge of the incumbent vendor. This approach – supporting competition while defending business principles such as intellectual property rights – keeps all parties honest, as well as helping vendors know where they stand at all times.

The second major factor we faced with the F404 was the influence of the incumbent vendor. G.E. is a smart, powerful corporation, with smart, powerful friends. Military procurement professionals need to be the equal of vendors, in both sophistication and influence on major decisions. Procurement practices need to be open and transparent, with bright lines identifying the authority (or absence thereof) of the procurement professionals.

CHAPTER 4

AEGIS – HELL HATH NO FURY…

BUSINESS PRINCIPLES

Major programs, and especially technology development programs, take place over a time span that encompasses many political cycles. Such programs must be managed for stability. NAVSEA oversaw and integrated all aspects of Aegis (ships, radar systems, missiles, etc.) allowing them to manage the complex interactions between the multitude of programs. This sped overall development, and reduced the costs from change orders and inter-vendor conflicts.

Technology development programs also benefit powerfully from realistic simulations. Admiral Meyer's use of the civilian air traffic system was extremely valuable, yet cost very little. He learned things that would have cost a great deal of time and money to incorporate had they been discovered later in the development cycle.

CHAPTER 5

THE SILENT SERVICE -
"UNDERHAND, UNDERWATER AND DAMNED UN-ENGLISH"

BUSINESS PRINCIPLES

Managing the evolution of technology over multiple generations of products requires long-term focus and management accountability. This means that people need to stay in place in their jobs for a long time. This prevents them from escaping from problems they create, and allows them to develop the expertise needed to avoid creating them in the first place. There is nothing wrong with promoting an officer in the same job several times, just so long as he serves our country well.

CHAPTER 6

SMART BOMBS NEED SMART BUYERS

BUSINESS PRINCIPLES

Allowing mid-level and junior people to drive ideas can work very well. Munitions are precisely the type of procurement that these more junior teams can excel with. The results go beyond lower costs to taxpayers. With cheaper ammunition, the warfighters can use more ammunition both in training and in combat, increasing effectiveness where it really counts. The China Lake team that worked on the "seeker" components of the HARM AGM-88 (high-speed anti radiation missile) is a great example. The more junior people had great ideas; we turned them loose and they got things done.

Capital goods and consumables both have potential for cost savings, but need to be managed differently. The technology behind capital goods generally changes meaningfully between orders. For example, when a new ship contract is let, it is intended to add new capabilities to the fleet, not just to add more of the same. With munitions (smart or otherwise), the technology often does not change at all between contracts. When it does change, it is generally incremental rather than dramatic. This allows procurement professionals to bring different sources of supply into more direct competition. During one production run, which may last for months or years, the purchaser has time to collect vendor performance and product performance data to use as the basis for educating other competitors. As most commercial purchasing professionals know, this alone can be enough to motivate the incumbent vendor to do his best to improve price and performance on subsequent contracts.

We must use distinct practices for distinct phases of a product's life cycle. During research and development, the Navy and Contractor system designers need to work collaboratively. As research progresses, both sides need to openly discuss how to create the most effective system possible. This has to be relatively separate from cost discussions, or people will become too cautious in conversation to accomplish meaningful dialogue, ultimately creating a higher cost anyway. Once a weapon has been developed and is heading for production, the existing Navy team is likely then to be the wrong one to manage actual procurement. Simply put, by that stage in the procurement process the buyers and sellers will be too close to one another to be effective. This is the type of situation that can make military personnel into advocates for the vendor (via simple human nature) rather than advocates of the taxpayers.

CHAPTER 7

HONOR ABIDES HERE –
GETTYSBURG AND OTHER LESSONS IN HOMELAND DEFENSE

BUSINESS PRINCIPLES

Program effectiveness depends on how precisely dollars spent match the specifics of the situation. Spending large amounts of money, with great speed and large headlines, is not useful if the wrong things are purchased. Moreover, as the situation becomes more and more removed from prior experience, greater care is needed to ensure that the program and related spending precisely match the new issues. This is true whether the issue is business use of the internet or terrorist threats to homeland security. Any organization given a new mission will rely on pre-existing methods, especially when under time pressures. We must methodically ensure a tight logical connection between major expenditures and the new factors impacting them. Otherwise we risk making honest mistakes that deflate the value of our efforts.

In the case of homeland security response, the problems are very different than what we have traditionally expected our first responders to deal with. The new priorities include reacting to massive events, readiness for NBC (nuclear, biological, chemical) contamination, and co-coordinating with numerous and extensive other responding parties. Likewise, homeland security prevention has its own unique new priorities. We should not expect each responding or prevention organization to individually develop the new planning capabilities that a single, centralized procurement authority can provide.

Centralized procurement will give our Homeland Security the integration, co-ordination, and purchasing power that we need. By providing tight oversight, it will ensure that the spending fits tightly with the new needs. A single authority will also be able to achieve vendor pricing and responsiveness far better than smaller purchasing groups.

CHAPTER 8

DAMN THE TORPEDOES: CONGRESS, THE SENATE AND PROCUREMENT

BUSINESS PRINCIPLES

The basic principles are timeless, dating to 1815 and before. Professional management is both more effective and more transparent. Oversight, from a Board of Directors or a Congressional committee, works better with managers that are stable, focused, and subject to regular scrutiny.

Senior management needs to be in place long enough for them to learn their jobs, for their tenure to be critical to their own careers, and for overseers to learn to work with them. They must also be powerful enough in the organization to force changes in the organization when needed. The guiding policies and procedures need to be well known and closely followed. When decisions go against these policies, or are otherwise important, they must be thoroughly scrutinized. These decisions need to be well documented and based on hard facts. This makes both procurement and oversight more effective.

CHAPTER 9

THE ENDLESS CRUSADE FOR SENSIBLE PROCUREMENT – TODAY'S BLUNDERS ARE TO-MORROW'S BILLS

BUSINESS PRINCIPLES

Professional purchasing gives more bang for the buck. Professional contracting (especially enforcement and cancellation) protects those bucks when things go wrong. New ideas and systems have risks. Some will be surprisingly effective; others will fail. Professional contracting allows us to proceed with the attractive ones without requiring prescience as to which will succeed. This is becoming much more important. The threats we face are becoming harder to predict; we will be surprised more often about what works and what doesn't. We must be ready to change directions.

Programs must have decision points and an exit strategy. There must be plans for how to salvage value (technology developed, etc.) that can benefit future programs. We need fair and explicit performance and cancellation terms. Otherwise, the cost of getting out of a program can be as expensive as building the system – except the costly cancellation does not add to our nation's defense. Non-performance terms must be specific, with allowance for cancellation at the major points where such decisions are typically taken. This is simple staged investing, a basic investment management technique employed by corporations and financiers around the world.

APPENDIX B

CHAPTER 10

CHARTING OUR COURSE TO THE FUTURE

BUSINESS PRINCIPLES

National defense is too important to be done in anything but a completely professional manner. Professional procurement management is one of the disciplines that private enterprise has rapidly and effectively adopted in recent years. For budgets as large as those in the Department of Defense – some of the largest purchasing budgets in the history of mankind – the net benefits of focusing the procurement agencies would be huge.

The central technique of professional procurement is centralized purchasing. The successful acquisition process used for the Joint Strike Fighter clearly demonstrates that central procurement management can provide for the needs of multiple services for new technologies, and there is simply no credible argument against central purchasing for older technologies. Along with centralized purchasing, professional procurement management requires high skill levels. For these skills to thrive, procurement must become a desirable career path in and of itself (without the ability to attract, develop and retain the best minds, the choice becomes one of centralized or decentralized mediocrity).

For centralized purchasing and professional grade skills to be effective, the entire program must have a high level of transparency. Though potentially an end in itself, it is absolutely necessary to have transparency in order for procurement management to have both substantial autonomy and effective oversight. Without autonomy, it loses capability; without oversight the loss is accountability.

This structure of autonomous professionals that know they will be held to account will then be able to make the most of the management techniques developed and perfected in the most competitive business environment ever: modern American industry.

CHAPTER 11
GREASING WHEELS ON THE BANDWAGON OF CHANGE
BUSINESS PRINCIPLES

At some point, business principles need to become hard recommendations. Currently, Defense acquisition falls under the realm of the Under Secretary for Defense (Acquisition and Technology) and is then divided among the services. Within the services it is further divided. For instance, the Navy has separate procurement offices for Ships, Aircraft, Space, and Supply.

A major procurement agency should be formed that brings the separate efforts of the services into a single operating unit. It should be headed by a four-star officer with the same rank as the CinCs. This Chief of the Defense Material Procurement Agency will serve a six-year term for the sake of continuity and to make the position attractive. The Chief should have the support of an executive order and be a military officer from any of the branches. The agency should focus on acquisitions with a face value of $250 million or more, the big ticket, long-term systems most vital to our defense.

The creation of this agency should be treated as nothing more complicated than any other large corporation revamping its buying department. U.S. companies have done this thousands of times over the last several years; there is no shortage of knowledge of how to do it right. This will mean the end to many old ways of doing things. It will mean better value, better results. For procurement, this is Transformation.

APPENDIX B

CHAPTER 12

THE FLOOD TIDE OF REARMAMENT

BUSINESS PRINCIPLES

The lessons learned come from both recent and ancient history. Sun Tzu, the Continental Congress, Winston Churchill, and Reagan's Department of the Navy all learned lessons that can help us today. Current American performance reflects both skilled expansions on those lessons and missed opportunities. The Navy and Air Force did a masterful job of managing competition for the JSF program. The Crusader program is another matter altogether.

We have the ability to adopt the private sector management tools that have been so effective in accelerating business and cutting costs, but we need leadership from the top to do so. Even without senior leadership, individuals within the system are developing examples of excellence that can be built upon. Air Force General Lester Lyles has adopted collaborative working with vendors and users, along with taking small steps instead of ponderously contracting for large ones. Both techniques were developed in commerce, especially for programs that depend on both.

Meanwhile, the external environment has deteriorated on multiple fronts. The supply base has become more concentrated. The immediacy of combat has taken the most powerful tool away from a weak purchaser – the ability to delay instead of accepting a bad deal. Finally, the supply base has increased pressure to boost profits. More bang for the buck begins with efficiency in management.

INDEX

A

A-12 deep-strike attack aircraft, 85–87
  cancellation of program for, 82
A-6 Intruder deep-strike aircraft, 85
AAMRAM AIM-120 air-to-air missile, 57
Aboulafia, Richard L., 87
Advanced Technology systems, 4
AEC. *See* Atomic Energy Commission (AEC)
Aegis (seaborne weapons system), 30
  cost of, 36
  on Spruance class destroyers, 37
Aegis-equipped ships, 140, 142
Afghanistan
  artillery unit not used in, 83n
AGS (advanced gun system), 99
Aim 7E-2 missile, 59
aircraft carrier(s), 124
  constructed in 1930s, 104–105
  construction of, 19
  CVNX-class of, 14
  fact sheet, 176–177
  future need for, 134–135
  home ports of, 16
  Nimitz class
    characteristics of, 15
    funding for, 16–17
    reactors for, 110
  planning for future needs for, 13–14
aircraft, military
  engines for, 23–24
  operational requirements of, 22
  procurement of fighter, 24–29
Aldridge, E.C. "Pete," 97
Alfred (American warship), 132–133
Alligator (American submarine), 42
Allison 501-K34 (GTGS), 34
Amendment #1516 (Levin-Hart), 16–17
Amphibious Assault Ships, 31
AN/SPY-1 (radar system), 32
  power requirements of, 34
Anderson, Spc.Marc, 100
Angrist, Gene, 86
Argo, 44
Argonaut I, 44
Argonaut Jr, 44
Arleigh Burke class destroyers (DDG), 30
  characteristics of, 33
  construction of, 37–38
  armaments contracts
  cancellation of, 82–84, 85–87
Army and Navy Club (Washington, D.C.), 128
artillery system(s), 83n
  AGS (advanced gun system), 99
  Crusader, 82–84, 100
atomic bomb, 54
Atomic Energy Commission (AEC), 46
Avondale and Ingalls, 4

B

Barnard, Richard, 25
Bath Iron Works (Maine), 4, 38n, 118–119
battles
  Guam, 107
  Iwo Jima, 107
  Marianas Turkey Shoot, 107
  Philippine Sea, 107
battleships
  constructed in 1930s, 104–105
  Iowa class, 108
  North Carolina class, 107
  South Dakota class, 108
Bethlehem Steel Company (Quincy, MA), 104
Bettis Atomic Power Laboratory, 46
Boeing, 61, 85
  and JSF production, 25

INDEX

growth of, 4
Bowman, Adm. "Skip," 47–48, 119–120
Bremerton, WA, naval shipyard. *See* Naval Shipyard (Bremerton, WA)
budget, defense 1983, 8–9
Busey, Adm. J.B., 25, 26
Bush administration
  rearmament policies of, 79–81
Bush, George W., 79–81
  on War on Terrorism, 132
Bush, George, Sr, 15

C

CAG. *See* Competition Advocate General of the Navy (CAG)
capital goods procurement, 7, 14, 15, 185
Carlucci, Frank, 17
Chamberlain, Joshua, 64–65
Cheney, Dick, 85–86
China
  as emerging world power, 90, 117, 135–138
  military capability of, 135–137
Churchill, Winston S., 102–104, 123–124
Citadel, The, 79
Clark, Adm. Vern, 98, 180
Cleveland, Grover, 116
Coastguard, U.S., 143
Cohen, Adm. Jay M., 98–99
Cohen, Harold, 86–87
Commons, Pfc. Matt, 100
communications
  homeland defense role of, 69–70
  in homeland defense, 65–66, 69–70
Competition Advocate General of the Navy (CAG)
  mandate of, 8
  Platt's appointment as, 2, 5–6

competition in procurement practice, 2–5, 181–183
congressional oversight, 187, 195
contract management, 2–7, 9, 17–26, 47, 52, 69, 79
  and patent rights, 43
  and the A-12 cancellation, 85–87
  cancellation penalties in, 83
  exit strategies in, 83–84
Cowley, Adm. Robert E., III, 120–121
Crose, Sgt. Brad, 100
cruisers, Ticonderoga class, 30, 32
Crusader artillery system, 82–84, 100, 117–118
competition management in procuring, 199
Cunningham, "Duke," 59
Cunningham, Sr Airman Jason D., 100
CVN 77 (US navy vessel), 14
CVNX class aircraft carriers, 14

D

Dayton-Wright Aircraft Company, 81
DD 21, 83
DD(X) ships, 97–100
DDG 51 class destroyers, 118–119
defense contractors, American
  mergers of, 3–5, 62
  problems of in 21st century, 124–128
Defense Logistics Agency (DLA), 109
Defense Material Procurement Agency (DMPA), 121, 198–199
  proposed by S. Platt, 113–114
defense spending, world comparison, 172–175
Department of Defense
  budget of compared to corporate revenues, 92

201

THE ARMAMENT TIDE

Navy budget for 2003, 96–97
  salaries of compared to corporations, 93–94
  spending of compared to Gross Domestic Product, 92–93
destroyers
  Arleigh Burke class, 30, 33
  construction of, 37–38
  DDG 51 class of, 118–119
  Spruance class of, 31
Dicks, Norm, 18
Directed Energy Weapons, 100

E
Eisenhower Award Dinner, 88
Electric Boat, Division of General Dynamics, 10, 36, 43, 49, 51
Ellis Island, 131n
engine(s), gas turbine
  G.E. LM2500 series, 38, 99
engine(s), gasoline
  in submarines, 43
engine(s), jet, 191
engine(s), steam
  in submarines, 42
Enola Gay (bomber), 54
ERT (Emergency Response Team), 67
Executive Order 12344 (February 1, 1982), 112–113
EXOCET (stand-off missile), 22
Exxon Mobil revenues, 92

F
F-35 JSF (Joint Strike Fighter), 24
F/A-18 Hornet, 22, 26, 27n
F/A-18 Super Hornet, 121
F/A-18E Super Hornet, 23
F/A-18F Super Hornet, 23
F404, 191–192
Falklands War, 21
Fenian Ram (submarine), 42

Fenian Society, 42
firefighters' role in homeland defense, 67
flag, Grand Union, 132–133
Fleischer, Ari, 116
Ford Motors revenues, 92
Ford, William Clay, Jr
  salary of, 93
Fowler, Earl, 18–19

G
Gas Turbine Generator Sets (GTGS), 34
GE F404 engine(s), 23, 25, 26, 27
  produced by Pratt and Whitney, 28–29
GE F414 engine(s), 23
General Dynamics, 10, 49, 62, 85–86, 87, 118
  growth of, 4
General Dynamics, Electric Boat Division. *See* Electric Boat, Division of General Dynamics
General Electric, 47, 191
  FBI investigation of, 25–26
  monopoly on aircraft engines by, 23–24
  turbine blade technology of, 26
General Motors, 62
  revenues of, 92
Gibraltar, Straits of, 31
Gilmer, Thomas, 53, 54
Grant, Ulysses S., 70–71, 116
Greenspan, Alan, 126
Guam, Battle of, 107
Gulfstream Aerospace, 4
gun(s)
  chrome-plated barrels for, 60
  Crusader, 83–84, 117
  M61-A1 Gatling Gun, 60
  rifled barrels of, 54
  See Also artillery systems

INDEX

H
H-3 (submarine), 44–45
Harding, Warren G., 116
HARM AGM-88 (high-speed anti-radiation missile), 61, 192
Harpoon AGM-84 (missile), 61
Hart, Gary
  amendment to reduce navy funding, 16–17
Havana Harbor
  explosion of USS *Maine* in, 128–130
Have Nap AGM-142 (Raptor missile), 62
Henry, Brig. Gen. Charles R., 69–70
Heseltine, Michael, 41
Historic Ships Foundation (San Francisco), 56–57
HMS *King George V*, 105–106
HMS *Punjabi*, 105–106
Hodges, Robert, Jr, 87
Holland class submarines
  characteristics of, 51
Holland I (British Navy submarine), 41
Holland IV (submarine), 42
Holland Torpedo Boat Company, 51
Holland, John, 42–43, 52
homeland defense, 194–195
  border security, 66
  centralized procurement authority in, 68–69, 186–187
  communications in, 69–70
  emergency response in, 65–67
  focus of, 180–181
  intelligence gathering, 66
  procurement for, 186–187
Honigman, Steven S., 70, 86
Hood, E.E., 27, 28
Hughes Aircraft Co, 61–62

I
immigration to U.S., 131
India, Chinese threat to, 136
Ingalls and Avondale shipyards, 118
Ingalls Shipyard (Pascagoula, Mississippi), 31, 36, 38n
Inhofe, James, 84
Integrated Power System (IPS), 99, 100
intellectual property
  ownership of, 43–44
intelligence gathering and analysis, 66, 67–68
interest rates
  effect of on defense procurement, 126
International Trade Center, 88
Iowa class battleships, 104–105, 108
Iowa Jima, Battle of, 107
Iran, 132
Iraq, 117–118, 132
Irish Republican Army, 42

J
Japan, Chinese threat to, 136
Jason, the Argonaut, 44
JASSM (joint air-to-surface stand-off missile), 97n
JDAM (joint direct attack munition), 97n
Johnson, Lyndon B., 110
Joint Strike Fighter (JSF), 23, 97, 121
  acquisition process for, 197
  competition management in procuring, 199
  production of, 24–25
Jones, Gen. James L., 63*ff*, 144
Jones, John Paul, 132–133
Joyner, Lt. Cmdr. Sara, 13
JSOW (joint stand-off weapon), 97n

203

K

Kennedy, John F., 110
Kettering "Bug," 63*ff*
Kettering, Charles F., 81
Kidd, Adm. I.C., 36, 37
Kilo class submarines (Russian), 136
Kirishima (Japan), 106
Kirk, Mark Steven, 137–138
Kissinger, Henry, 94–95
Knight, Cmdr. Russ, 13
Krebs, James (General Electric), 25, 27

L

Laird, Melvin, 95
Lake, Simon, 43, 44
Lehman, John F., Jr, 2, 6, 10, 17n, 18, 28, 48
Leven, Peter, 41
Levin, Senator
  amendment to reduce navy funding, 16–17
LGB (laser guided bomb), 97n
LHA(s) (Amphibious Assault Ships), 31
Little Boy (atomic bomb), 54
Little Round Top, Battle of, 65
Litton Industries, 4
LM2500 gas turbine engines, 99
Lockheed Martin
  and JSF production, 24–25
  growth of, 4–5
Logicon, 4
Loral, 4
Los Angeles class attack submarine(s), 7–8, 48, 52
LPD 17 class amphibious assault ships, 118
LPD 19 class assault ships, 118–119
Lyles, Gen. Lester, 122, 199

M

M61-A1 Gatling Gun, 60
Macarthur, Gen. Douglas, 121
Marianas Turkey Shoot, Battle of, 107
Martin Marietta, 4
Mcdonnell Douglas Corporation, 4, 61, 85–86, 87
McKee, Kinnaird, 48
McNamara, Robert, 109–110, 114–115
Meade, Gen. George, 70
media, American
  role of in creating terror, 133
mergers, corporate
  effect of on competition, 125–126
Meyer, Wayne, 33–34, 142
Michaelis, Adm. Frederick Hayes, 27n, 37
military personnel
  recruitment and retention of personnel, 94–95
missile(s), 57–59
  AIM 7E-2, 59
  cost of, 61
  HARM AGM-88 (high-speed anti-radiation missile), 193
  surface-to-air, 58–59
  Trident, 63*ff*
  types of, 58, 61–62, 62, 97n
MK 48 (torpedo), 61
MK 50 (torpedo), 61
Morris, William, 87
MOU. *SEE* Navy Memorandum of Understanding (MOU)
Myers, Richard B., 88

N

National Guard, 143
  homeland defense role of, 67, 68, 181
Natter, Adm. Bob, 144
Natter, Claudia, 144

INDEX

Naval Nuclear Propulsion Program, 49, 50, 111–113, 119
Naval Shipyard (Bremerton, WA), 18, 45, 106, 107
Naval Weapons Center (China Lake, CA), 61
Navy Memorandum of Understanding (MOU), 118
Navy, U.S.
 budget of for 2003, 96–97
New York Naval Shipyard, 104
New York Shipbuilding Corp. (Camden, NJ), 104
Newport News Shipbuilding, 4, 18, 36
Nixon World Enterprises, Inc, 137
Nixon, Edward, 137
Nixon, Richard, 94–95
North Carolina class battleships, 105, 107
North Korea, 117–118, 132
Northern Command (NORTH COM), 67
Northrop Grumman, 4–5
Northrop Grumman Shipbuilding Services, 118
nuclear power
 decline numbers of strategic war heads, 51–52

O

O'Connor, John, 55–56
Office Naval Research, 60
Ohio class submarines
 characteristics of, 51
Oklahoma City bombing, 1995
 communications failure during, 65–66
Old Ebbitt Grill (Washington, D.C.), 116, 117
Operation Enduring Freedom, 134
ordnance. *SEE* weapons
Orpheus, 44

P

Page, Susan, 134n
patent rights, 43
Pei Cobb Freed and Partners (New York Architects), 88
Penguin AGM-119B (anti-ship missile), 62
People's Liberation Army, China (PLA), 136–137
Persian Gulf War, 15
PGMs (precision guided munitions), 97
Phalanx Close-In Weapons System, 60
Philadelphia Navy Yard, 53, 105
Philippine Sea, Battle Of, 107
Philippines, The, 132
Phoenix AIM-54C (air-to-air missile), 61
Pier 37 (Seattle, WA), 140
Plank Owners, 141–142
Platt, Stuart F.
 appointed Competition Advocate General of the Navy (CAG), 2, 5
Plunger (American submarine), 42
Plunger series submarines, 44
police forces
 homeland defense role of, 67
political appointees
 recruitment and retention of, 109
population of U.S., 131
Pratt and Whitney, 23–24
 and production of GE F404, 28–29
procurement practice
 capital goods acquisition in, 193–194
 congressional oversight in, 187, 195
 for capital good acquisition, 7, 14, 15, 185
 for capital goods purchases, 185
 for small ticket items, 60, 185

205

personnel for, 184
recruitment and retention of
  personnel for, 188
research and development in, 47,
  63, 97, 184, 194
stability management in, 183
See Also sourcing in procurement
  practice
procurement, military, 2
  budgets for rising, 96
  centralized authority for, 68–69,
    91–92, 103, 113, 187,
    197
  competition in, 188–191
  government policies on, 25n,
    180–181
  inefficiencies of multiple offices
    for, 91
  recruitment and retention of
    personnel for, 94–95, 188
Project on Government Oversight,
  86
Puget Sound Navy Yard
  (Bremerton, WA).
  See Naval Shipyard
  (Bremerton, WA)
Pulse Weapons, 100
Pyatt, Ev, 26, 53

R
radar system (AN/SPY-1), 32
Raptor (missile), 62
Raymond, Lee R. (Exxon Mobil)
  salary of, 93
Raytheon, 61–62
reactor(s)
  for Nimitz class carriers, 110
  naming of, 46n
  nuclear, 49
  S5W, 46
  S6G, 46
Reagan administration
  rearmament policies of, 1–2, 6
Reagan, Ronald, 2, 6, 111

rearmament
  management of, 138–140
Reimer, Gen. Dennis, 64
Renfro, Adm. E.E., III, 36, 37
Reno, Janet, 5n–6
research and development, 194
  in procurement practice, 184
  stability management of, 192
research and development in
  procurement practice,
  47, 63, 97
Reserve Forces' role in homeland
  defense, 68, 181
Rice, Condoleezza, 88
Richardson, Elliot L., 110
Rickover, Adm. Hyman George,
  10, 19–20, 34–36, 45,
  46, 48–49, 98, 110, 111
  and USS *Maine* investigation,
  129–130
Rickover, Eleanore, 48–49
Rickover, Ruth Masters, 49n
Ridge, Tom, 66
Roberts, Pfc Neil, 101
Roosevelt, Franklin Delano, 17,
  53, 124n
Roosevelt, Theodore, 17, 44, 53,
  116
Rota, Spain, 31, 45
Roth, William V., 21–22
Rumsfeld, Donald, 79, 82, 84, 88,
  99
  salary of, 93
Russia
  superpower status of, 1

S
S5W reactor, 46
S6G reactor, 46
Sambur, Marvin, 122
SAMs (surface-to-air missiles),
  58–59
Sawyer, George, 17, 53
Scapa Flow, U.K., 105

INDEX

Schlesinger, James, 95
Scott, H. Lee (Wal-Mart)
  salary of, 93
Seattle Space Needle, 140
Seawolf attack submarine(s), 7–8
  construction program for cancelled, 52, 82
Shaara, Michael, 64
Shalikashvili, Gen. John M., 144
shareholder expectations
  effect of on defense procurement, 126
Shinseki, Eric K., 84
ships, British. *See* names of individual ships beginning with HMS
ships, Japanese Navy
  Kirishima battleship, 106
ships, U.S. Navy
  Aegis-equipped, 140, 142
  class designation of, 38n
  commissioning of, 141–142
  congressional debate over funding for, 16–17
  construction programs for in 1930s, 104–105
  cost of, 7
  DD(X), 97
  LPD 17 class amphibious assault, 118
  LPD 19 class assault, 118–119
  number and type of, 8, 9
  planning for future needs for, 13–15, 14–15
  See Also names of individual vessels beginning with USS
Shoup, Gen. David Monroe, 140–141
Sidewinder (missile), 61
Sigsbee, Capt. Charles, 128
Sixth Fleet, 107
SLAM-ER (stand-off land attack missile), 97n
Somalia, 132

Song class guided missile submarine, Chinese Navy, 136
sourcing in procurement practice, 18, 23–25, 43, 60–61, 66, 111, 125, 139
South Dakota class battleships, 108
Sparrow (missile), 61
Spruance class destroyers, 31
  transition to Ticonderoga class, 37
Spruance, Adm. Raymond A., 107
stability management in procurement practice, 183
Starr, Barbara, 116n
Sturgeon class of submarines, 46
submarine(s)
  Chinese Song class, 136
  design of, 41–42
  Holland class, 51
  Kilo class (Russian), 136
  Los Angeles class, 7–8, 52
  nuclear, 31
  Ohio class, 10, 51
  Seawolf class, 52
  SSBN 598/608 class, 31
  SSBN 640 class Poseidon, 31
  Sturgeon class of, 46
  Virginia class, 52
Svitak, Sgt. Philip J., 100
Syria, 132

T

Taft, William Howard, IV, 27, 28n
Taiwan, Chinese threat to, 136
Task Force 39 (Scapa Flow, U.K.), 105
Taylor, Elizabeth, 35
Teal Group, 87
technology, military, 96
  ownership of, 26–27
  rapid advances in, 125
  research and development of, 192
  stability management of, 192
terrorism

cyber, 133
economic, 133, 134
psychological, 133
See Also homeland defense
Terrorism, War on, 132–135, 140
Thatcher, Margaret, 41n
Thompson, Mark, 7
Ticonderoga class cruisers (CG), 30
 characteristics of, 32
Tolstoy, Leo, 123
Tomahawk Cruise Missile, 97n
Trident missiles, 63*ff*
Trident submarines, 10
Truman, Harry S, 121n
TRW, 4

U

United Technologies (Hartford, CT), 24
unmanned aerial vehicle (UAV), 81–82
USA Freedom Corps
 homeland defense role of, 68, 181
USAF F-16, 23
USS A-1 (SS2), 44
USS *Abraham Lincoln* (CVN 72), 16, 17
USS *Arleigh Burke* (DDG 51), 38–39
USS *Bainbridge* (CGN 25), 46n
USS *Boone*, 63*ff*
USS *Bunker Hill* (CG 52), 37
USS *Cole* (DDG 67)
 terrorist attack on, 38–39
USS *Enterprise* (CVN 65), 18, 27n
USS *George Washington* (CVN 73), 16, 17
USS *Holland* (AS 32), 45
USS *Holland* (SS1), 43
USS *Hyman G. Rickover* (SSN 709), 48, 98–99
USS *Indiana* (BB 58), 106

USS *Iowa* (BB 61), 56–57, 63*ff*, 104–105
USS *John C. Stennis* (CVN 74), 13–16
USS *John F. Kennedy* (CV 67), 110
USS *Kittyhawk* (CV 63), 14
USS *Maine*, 128–130, 140
USS *Massachusetts* (BB 59), 104
USS *Milwaukee* (C-21), 44–45
USS *Nautilus* (SSN 571), 10
USS *North Carolina* (BB 55), 104
USS *Norton Sound* (AVM 1), 32
USS *Plunger*, 44
USS *Princeton*, 53
USS *Ranger* (CV 4), 104
USS *Ronald Reagan* (CVN 76), 14
USS *Shoup* (DDG 86), 140–143
 cost of, 143
USS *Sigsbee* (DD 502), 140
USS *Simon Lake* (AS 33), 31, 43, 45
USS *South Dakota* (BB 57), 104
USS *Ticonderoga* (CG 47), 32
USS *Underwood* (FFG 36), 60
USS *Waldron* (DD 699), 55–56
USS *Washington* (BB 56), 105–108
USS *Winston Churchill* (DDG 81), 38

V

V22 Osprey VTOL (Vertical Take Off and Landing), 97
Vance, SSgt Kevin, 100–101
Veliotis, Takis, 49
Verne, Jules, 44
Vicksburg, Battle of, 70
Virginia class submarines, 52
VLS (Vertical Launching System), 37–38

INDEX

W
Wagoner, G. Richard, Jr (General Motors)
  salary of, 93
Wal-Mart
  revenues of, 92
Walker, Sgt., 101
Warner, John, 35
weapons, 53–63
  atomic bomb, 54
  cost of, 55
  Directed Energy, 100
  Pulse, 100
  types of, 97n
  See Also missiles
Weinberger, Caspar W., 2, 48, 94–95
Westinghouse Electrical Systems, 4, 47
Wilcox, Adm. John W., 105
Willard Hotel, The, 88
Wilson, A.K., 40, 41, 114–115
World Trade Center attacks, 2001
  communications failure during, 66

Y
Yemen, 132

Z
Zumwalt, Admiral, 35